ESS Emacs Speaks Statistics

A catalogue record for this book is available from the Hong Kong Public Libraries.

Published in Hong Kong by Samurai Media Limited.

Email: info@samuraimedia.org

ISBN 978-988-8381-82-1

Table of Contents

12   Overview of ESS features for the S family

13   ESS for SAS **68**

14   ESS for BUGS **77**

15   ESS for JAGS **78**

16   Bugs and Bug Reporting, Mailing Lists **80**

Appendix A   Customizing ESS **82**

Indices **83**

1 Introduction to ESS

The S family (S, Splus and R) and SAS statistical analysis packages provide sophisticated statistical and graphical routines for manipulating data. **E**macs **S**peaks **S**tatistics (ESS) is based on the merger of two pre-cursors, S-mode and SAS-mode, which provided support for the S family and SAS respectively. Later on, Stata-mode was also incorporated.

ESS provides a common, generic, and useful interface, through emacs, to many statistical packages. It currently supports the S family, SAS, BUGS/JAGS, and Stata with the level of support roughly in that order.

A bit of notation before we begin. *emacs* refers to both *GNU Emacs* by the Free Software Foundation, as well as *XEmacs* by the XEmacs Project. The emacs major mode `ESS[language]`, where `language` can take values such as `S`, `SAS`, or `XLS`. The inferior process interface (the connection between emacs and the running process) referred to as inferior ESS (`iESS`), is denoted in the modeline by `ESS[dialect]`, where `dialect` can take values such as `S3`, `S4`, `S+3`, `S+4`, `S+5`, `S+6`, `S+7`, `R`, `XLS`, `VST`, `SAS`.

Currently, the documentation contains many references to 'S' where actually any supported (statistics) language is meant, i.e., 'S' could also mean 'R' or 'SAS'.

For exclusively interactive users of S, ESS provides a number of features to make life easier. There is an easy to use command history mechanism, including a quick prefix-search history. To reduce typing, command-line completion is provided for all S objects and "hot keys" are provided for common S function calls. Help files are easily accessible, and a paging mechanism is provided to view them. Finally, an incidental (but very useful) side-effect of ESS is that a transcript of your session is kept for later saving or editing.

No special knowledge of Emacs is necessary when using S interactively under ESS.

For those that use S in the typical edit–test–revise cycle when programming S functions, ESS provides for editing of S functions in Emacs edit buffers. Unlike the typical use of S where the editor is restarted every time an object is edited, ESS uses the current Emacs session for editing. In practical terms, this means that you can edit more than one function at once, and that the ESS process is still available for use while editing. Error checking is performed on functions loaded back into S, and a mechanism to jump directly to the error is provided. ESS also provides for maintaining text versions of your S functions in specified source directories.

1.1 Why should I use ESS?

Statistical packages are powerful software systems for manipulating and analyzing data, but their user interfaces often leave something something to be desired: they offer weak editor functionality and they differ among themselves so markedly that you have to re-learn how to do those things for each package. ESS is a package which is designed to make editing and interacting with statistical packages more uniform, user-friendly and give you the power of emacs as well.

1.1.1 Features Overview

- Languages Supported:
  - S family (R and S+ AKA S-PLUS)

- SAS
- OpenBUGS/JAGS
- Stata
- Julia
- Editing source code (S family, SAS, OpenBUGS/JAGS, Stata, Julia)
  - Syntactic indentation and highlighting of source code
  - Partial evaluation of code
  - Loading and error-checking of code
  - Source code revision maintenance
  - Batch execution (SAS, OpenBUGS/JAGS)
  - Use of imenu to provide links to appropriate functions
- Interacting with the process (S family, SAS, Stata, Julia)
  - Command-line editing
  - Searchable Command history
  - Command-line completion of S family object names and file names
  - Quick access to object lists and search lists
  - Transcript recording
  - Interface to the help system
- Transcript manipulation (S family, Stata)
  - Recording and saving transcript files
  - Manipulating and editing saved transcripts
  - Re-evaluating commands from transcript files
- Interaction with Help Pages and other Documentation (R)
  - Fast Navigation
  - Sending Examples to running ESS process.
  - Fast Transfer to Further Help Pages
- Help File Editing (R)
  - Syntactic indentation and highlighting of source code.
  - Sending Examples to running ESS process.
  - Previewing

ESS provides several features which make it easier to interact with the ESS process (a connection between your buffer and the statistical package which is waiting for you to input commands). These include:

- **Command-line editing** for fixing mistakes in commands before they are entered. The '-e' flag for S-plus provides something similar to this, but here you have the full range of Emacs commands rather than a limited subset. However, other packages do not necessarily have features like this built-in. See Section 4.1 [Command-line editing], page 26.

- **Searchable command history** for recalling previously-submitted commands. This provides all the features of the 'Splus -e' history mechanism, plus added features such as history searching. See Section 4.3 [Command History], page 28.

- **Command-line completion** of both object and file names for quick entry. This is similar to `tcsh`'s facility for filenames; here it also applies to object names and list components. See Chapter 9 [Completion], page 46.

- **Hot-keys** for quick entry of commonly-used commands in 'S' such as `objects()` and `search()`. See Section 4.5 [Hot keys], page 31.

- **Transcript recording** for a complete record of all the actions in an S session. See Section 4.2 [Transcript], page 26.

- **Interface to the help system**, with a specialized mode for viewing S help files. See Chapter 8 [Help], page 44.

If you commonly create or modify S functions, you will have found the standard facilities for this (the '`fix()`' function, for example) severely limiting. Using S's standard features, one can only edit one function at a time, and you can't continue to use S while editing. ESS corrects these problems by introducing the following features:

- **Object editing**. ESS allows you to edit more than one function simultaneously in dedicated Emacs buffers. The ESS process may continue to be used while functions are being edited. See Section 7.1 [Edit buffer], page 37.

- **A specialized editing mode** for S code, which provides syntactic indentation and high-lighting. See Section 7.4 [Indenting], page 38.

- **Facilities for loading and error-checking source files**, including a keystroke to jump straight to the position of an error in a source file. See Section 7.3 [Error Checking], page 38.

- **Source code revision maintenance**, which allows you to keep historic versions of S source files. See Section 7.6 [Source Files], page 40.

- **Facilities for evaluating S code** such as portions of source files, or line-by-line evaluation of files (useful for debugging). See Chapter 5 [Evaluating code], page 34.

Finally, ESS provides features for re-submitting commands from saved transcript files, including:

- **Evaluation of previously entered commands**, stripping away unnecessary prompts. See Section 4.2.3 [Transcript resubmit], page 27.

1.2 New features in ESS

Changes and New Features in 15.09:

- ESS[R]: The indentation logic has been refactored. It should be faster, more consistent and more flexible. There are three types of indentation settings, those starting with `ess-offset-` give the actual offsets, those starting with `ess-indent-` are control (commonly Boolean) variables, and those starting with `ess-align-` are vertical alignment overrides which inhibit default offsets in specific situations. See `ess-style-alist` for detailed description of the new indentation system and provided default indentation styles.

- ESS[R]: Deprecation of old indentation settings. As a consequence of the indentation re-factoring `ess-brace-imaginary-offset`, `ess-expression-offset` and all delimiter-specific offsets are deprecated. The settings for indentation of continued statements have been replaced by `ess-offset-continuations`. It can be set to either cascade or `straight` (the default). `ess-arg-function-offset` has been replaced by `ess-indent-from-lhs` and has been generalised to assignements. This setting now works with both statement blocks and expressions and only takes effect for offsets set to `prev-call` and `open-delim` in order to produce a consistent indentation.

- ESS: A test framework has been set up.

- ESS[R]: A new RStudio style is provided to mimic as closely as possible R files indented via RStudio. To reproduce the setup of some of the RStudio users, the RStudio- style with `ess-offset-arguments` set to `prev-line` is also provided. In addition, the new RRR+ style is equivalent to RRR except it indents blocks in function calls relatively to the opening delimiter. This style does not try to save horizontal space and produces more indentation.

- ESS[R]: Roxygen fields will now be indented on paragraph refilling in order to make the documentation more readable. You can also refill commented lines in the `examples` field without squashing the surrounding code in the comments.

- ESS[R]: ESS can now format your code! This is controlled through the settings `ess-fill-calls` and `ess-fill-continuations`. When activated, `(fill-paragraph)` formats your calls and your formulas/continuations while making sure they don't go past `fill-column`. Repeated refills cycle through different styles (see the docstrings for more details). By default, the refilled region blinks. Set `ess-blink-filling` to nil to prevent this.

- ESS[R]: Fix occasional missing error location fontification in inferior buffers.

- ESS[R]: ess-developer now correctly assigned the environment of new functions to the package namespace.

- ESS[Julia]: ?[topic] now works in the *julia* buffer. Note that support for editing Julia code now depends on `julia-mode.el` from the Julia project. If you install ESS from the official tarball/zip file, `julia-mode.el` is already included. Otherwise, if you install ESS by running `make`, then the latest version of `julia-mode.el` is downloaded (and so you need an active internet connection to install) during the installation process. Alternatively, if you run ESS without running `make`, then ensure that you have the `julia-mode.el`, which you can get easily from MELPA for example.

- iESS: For naming inferior processes, ESS can use `projectile`'s project root and it does so when `ess-gen-proc-buffer-name-function` is set to `ess-gen-proc-buffer-name:projectile-or-simple` as by default, or to another value beginning with `ess-gen-proc-buffer-name:projectile-*`.

Changes and New Features in 15.03-1:

- ESS[R]: An indentation bug has been fixed (github issue 163)

- ESS[R]: On windows, if 'ess-prefer-higher-bit' is non-nil (the default), then R-newest will try to run a 64 bit (rather than 32 bit) version of R.

Changes and New Features in 15.03:

- ESS[R]: Full native support for 'company-mode'.

- ESS[R]: More efficient caching algorithm for R completion.

- ESS: New offset variable 'ess-close-paren-offset' to control the indentation of the closing parentheses.

- ESS[R]: Ask for CRAN mirror only once per emacs session.
- ESS[R]: Detect `library` and `require` calls for better completion caching.
- Buffer display is now customizable (`ess-show-buffer-action`).
- Use `y-or-n-p` instead of `yes-or-no-p` throughout.
- More support for ODS in ess-sas-graph-view.
- Makefiles are now both UNIX and GNU friendly.
- ESS[R]: Simplify directory lookup in `ess-developer` (#137).
- Make closed paren indentation consistent

Bug Fixes in 15.03:

- Fix open brace indentation bug (#27 in ess/R-ESS-bugs.R).
- Fix git version lookup
- Don't check directory modtime in R dialect.
- Declare all ess macros for edebug.
- Add `ess-smart-comma` to eldoc message functions.
- Inform users when retrieving RDA aliases.
- Line ending in '~' is also a continuation line.
- Filing roxy paragraphs works as expected now.
- In `ess-developer-add-package`, remove incorrect 'wait' argument from `ess-get-words-from-vector` call.
- Fix #96, #117, #120, #125, #134, #137.
- Fix ess-quit-r. Call base::q() even if it is masked.
- Fix 'ess-show-buffer' to always display the buffer in another window.
- Makefile: Fix cd bug for directories with spaces in them
- `ess-kill-buffer-and-go` modified to not restart R

Changes / Selected Bug Fixes in 14.09:

- ESS[Julia]: Executable is changed to `julia`.
- ESS[Julia]: Completion and help system was adjusted to Julia v.0.3.0. Julia v.0.2.x is no more supported.
- ESS[R]: Running R with `gdb` debugger now works as expected
- iESS: Inferior ESS buffers are now derived from `comint-mode`
- ESS[R]: `ess-execute-screen-options` uses correct screen width in terminal sessions
- ESS: `ess-build-tags-for-directory` works when no TAGS file name was provided
- ESS: `ess-offset-statement-continued` is now respected everywhere except inside of the `if` test condition.
- ESS: New variable `ess-offset-statement-first-continued` for indentation of the first line in multiline statements.
- ESSR: Starting `,` in multiline statements indentation is now ignored to achieve a more pleasant alignment.
- ESSR: Improved behavior of *RET* in roxygen blocks.

- ESS[R]: command cleaning with `C-u C-u C-y` was broken with lines containing " + "
- ESS[R]: fixed "empty watch window bug"
- ESS[R]: don't ask for help location on ac-quick-help (request of github #81)
- ESS[R]: "importClassesFrom" and "importMethodsFrom" were added to the list of two-parameter roxygen commands
- ESS[R]: fix vignetes display and hyperlinks (were broken in 13.09-1)
- ESS[Julia]: recognize function names ending with !
- ESS[Julia]: fix indentation of "for" comprehension syntax within brackets.

Changes / Selected Bug Fixes in 13.09-1:

- ess-remote and TRAMP: R support code is now downloaded in binary form instead of being injected from local machine. The R code is stored in `~/.config/ESSR/` directory on the remote machine
- TRAMP: PAGER environment variable is now correctly set to `inferior-ess-pager`
- retrieval of help topics on remote machines is fixed
- org-babel: source references of R code executed from org files correctly point to source references in original org files (version 8.2.1 or higher of org-mode is required for this feature)
- `ess-execute` is now bound to `C-c C-e C-e` in `ess-extra-map`.
- completion works again in `ess-execute`
- ESS[R]: `head` and `tail` methods were replaced by `htsummary` in `ess-R-describe-object-at-point-commands`
- ESS[roxygen]: evaluation commands now work in roxygen blocks. Leading comments are automatically removed before the evaluation
- ESS[transcript]: 'Clean Region' now works with multiline statements; `ess-transcript-clean-region` etc. correctly treat multiline statements, i.e., no longer forgets the lines typically preceded by '+'
- ESS[SAS]: Three features/fixes with special thanks to Matthew Fidler https://github.com/emacs-ess/ESS/pulls/mlf176f2. Turn on SAS log mode when appropriate. Indent comments and CARDS statement more appropriately.
- ESS[SAS]: `ess-sas-edit-keys-toggle` default returns to `nil`
- ESS[R]: support for `prettify-symbols-mode`: contribution from Rudiger Sonderfeld https://github.com/emacs-ess/ESS/pull/65
- ESS[SWV]: knitr now evaluates in the current frame
- ESS[developer]: ess-developer doesn't kill open DESCRIPTION files anymore
- ESS[roxygen]: `ess-roxy-preview-HTML` is now on `C-c C-o C-w` and `ess-roxy-preview-text` is now on `C-c C-o C-t`
- ESS: installation with `make install` was simplified and should work out of the box on most *nix systems
- ESS installation instructions simplified
- fixed font-lock bug introduced in 13.09 that was causing very slow process output

Changes/New Features in 13.09:

- font-lock in process buffers doesn't "spill" over prompts. Missing closing string delimiters should not cause wrong fontification of the following command input.
- ESS[julia]: full features M-TAB completion and auto-complete support, which now works for modules, structures and data types.
- ESS[julia]: a much better eldoc showing arguments of methods and data type constructors
- ESS-developer:
  - ESS-developer work-flow pattern has been streamlined: ESS-developer is now automatically activated on per-file basis if the file is part of a developed package `ess-developer-packages`. The old behavior (activation on per-process basis) is still available on `M-x ess-developer` in a process buffer.
  - integration with `devtools` package. New command `ess-developer-load-package` calls `load_all` on the package containing current file. `ess-developer-add-package` now offers IDO menu completions with available loading methods, currently `library`, and `load_all`. Loading command can be customized with `ess-developer-load-on-add-commands`.
- *TAB* now indents region if region is active (a contribution of Matthew Fidler in pull #41)
- `M-x ess-version` now reports full loading path and recognizes git and ELPA versions.
- warning and error keyword are now highlighted with `font-lock-warning-face` as they should be, (for quite some time these keywords have been hijacked by compilation mode fontification).
- eldoc: Eldoc now recognizes multiple processes. If current process is busy, or current buffer is not associated with a process, eldoc picks its completions from the first available free process.
- org-babel: evaluation is now org-friendly
- help: new help buffers now try to reuse ess-help buffers. This behavior is controlled by `ess-help-reuse-window` custom variable.
- help: ?foo pops IDO menu on multiple help files (so far it worked only for *C-c C-v*)
- remote evaluation is considerably faster now on slow connections
- ESS[R] tracebug R source references regular expressions are (mostly) language agnostic.
- `ess-function-call-face` inherits from `font-lock-function-name-face` rather than `font-lock-builtin-face`.
- `ess-inject-source` now accepts `function-and-buffer` option.
- Documentation: The "New Features" section (and `NEWS`) now represent recent changes: within the last year or so. All changes can be found in the new news.html (or `NEWS` and `ONEWS`).
- ESS[R] `ess-rep-regexp` should no longer inf.loop (rarely!), and hence `M-x ess-fix-miscellaneous` should neither.

Changes/New Features in 13.05:

- ESS[gretl]: Support for `gretl` (both editing and sub-process interaction). A contribution of Ahmadou Dicko.

- ESS: process output display is 4-10 times faster due to new caching and only occasional emacs re-display (for the moment this functionality is available only when `ess-tracebug` is active).

- ESS: `C-c '` is now bound to `ess-show-traceback` and `C-c ~` is bound to `ess-show-call-stack`.

- ESS[R]: ESS stores function in 'ESSR' environment to avoid kludging users' global environment and accidental deletion.

- ESS[R]: new variable `ess-swv-processing-command` to control weaving and tangling.

- ESS[R]: `ess-default-style` has been changed (from `DEFAULT`) to `RRR`. Use something like `(setq ess-default-style 'DEFAULT)` or `(setq ess-indent-level 2)` in your `~/.emacs` equivalent *before* loading ESS, if you do not like this new "incompatible" default style.

- ESS[julia]: ESS stores its functions in 'ESS' module.

- ESS[julia]: Eldoc is now supported in julia modes

- ESS[julia]: Adjusted error reference detection and interactive help to julia internal changes

- ESS[R]: `ess-use-tracebug`'s default has been changed to `t`. Set it to nil if you want to keep the previous behavior.

- ESS[tracebug]: Electric debug keys have been removed [breaking change] The functionality was replaced with `ess-debug-minor-mode` and `ess-debug-minor-mode-map`.

- ESS[tracebug]: `ess-tracebug-map` is an alias to `ess-dev-map` `C-c C-t`.

- ESS[tracebug]: `ess-bp-toggle-state` (`C-c C-t o`) can now be used during the debug session to toggle breakpoints on the fly (suggestion by Ross Boylan).

- ESS[tracebug]: `ess-debug-flag-for-debugging` and `ess-debug-unflag-for-debugging` work correctly from the debugging contexts. These commands also recognize non-exported functions for the packages listed in `ess-developer-packages` (`C-c C-t C-a`).

- ESS[R]: Eldoc (activated by `ess-use-eldoc`) has become more sophisticated, and hence also more intruding in the interface between the Statistics softare, e.g., R, and the user.

  Note that you can turn off ElDoc, by placing `(setq ess-use-eldoc nil)` in your `~/.emacs` file, prior to loading ESS,

- ESS[SAS]: long over-looked `SAS-mode-hook` appears!

- ESS[SAS]: `ess-sas-edit-keys-toggle` now defaults to `t` since `sas-indent-line` is still broken, i.e. *TAB* is now bound to `ess-sas-tab-to-tab-stop` by default

Changes/Bug Fixes in 12.09-2:

- ESS: new `ess-switch-to-end-of-proc-buffer` variable that controls whether `C-c C-z` switches to the end of process buffer. The default is `t`. You can use prefix argument to `C-c C-z` to toggle this variable.

- ESS: fix in `ess-eval-linewise` that was causing emacs to hang during R debugging with `ess-eval-visibly` equal to `t`.

- ESS: fix in `ess-eval-linewise` that was causing emacs to recenter the prompt in visible window

- ESS[tracebug]: A better handling of "Selection" prompts and debug related singlekey commands.

- ESS: fix a bug in `ess-switch-process` that was causing `*new*` selection to fail.

- ESS[R]: Solve missing `ess-local-process-name` bug in R-dired.

- ESS[SWV]: `ess-swv-PDF` doesn't ask for a command to run if there is only one command in `ess-swv-pdflatex-commands`.

- ESS[SWV]: `ess-swv-weave` gained an universal argument to allow for an interactive choice between available weavers (sweave, knitr).

- ESS: `ess-eval-*-and-step` functions go to next empty line at eob, instead of staying at the last line.

Changes/New Features in 12.09-1:

- ESS *Breaking Changes in Keys*:
    - New keymaps: `ess-doc-map` bound to *C-c C-d*; `ess-extra-map` bound to *C-c C-e*; `ess-dump-object-into-edit-buffer` was moved on *C-c C-e C-d*
    - roxygen map was moved on *C-c C-o* and *ess-roxy-update-entry* now resides on *C-c C-o C-o*
    - ess-handy-commands is not bound anymore
    - `ess-dev-map` (including `ess-tracebug` and `ess-developer`) moved on *C-c C-t*
    - *C-c C-y* is deprecated in favor of *C-c C-z C-z*

- ESS[R] new command `ess-describe-object-at-point` bound to *C-c C-d C-e* (repeat *C-e* or *e* to cycle). It was inspired by Erik Iverson's `ess-R-object-tooltip`. Customize `ess-describe-at-point-method` to use tooltip instead of an electric buffer.

- ESS: New command `ess-build-tags-for-directory` bound to *C-c C-e C-t* for building dialect specific tag tables. After building tags use *M-.* to navigate to function and objects definitions. By default *C-c C-e C-t* builds tags based on imenu regular expressions and also include other common languages `.c`, `.o`, `.cpp` etc. But it relies on external `find` and `etags` commands. If `ess-build-tags-command` is defined (for `R`), the inferior process is asked to build tags instead.

- ESS: `ess-switch-process` offers `*new*` alternative to start a new process instead of switching to one of the currently running processes.

- ESS: Switching between processes (*C-c C-s*) uses buffer names instead of the internal process names. Use `M-x rename-buffer` command to conveniently rename your process buffers.

- ESS: Process buffers can be automatically named on process creation according to user specified scheme. Default schemes are *proc*, *proc:dir* and *proc:abbr-long-dir* where `proc` stands for the internal process name and `dir` stands for the directory where the process was started in. The default is *proc*. For customization see `ess-gen-proc-buffer-name-function`.

- ESS: `ess-eval-visibly-p` is deprecated in favor of `ess-eval-visibly`.

- ESS: New evaluation pattern `nowait`. In addition to old `nil` and `t` values, `ess-eval-visibly` accepts `nowait` for a visible evaluation with no waiting for the process. See `ess-eval-visibly` for details on evaluation patterns.

- ESS: New "Process" menu entry with process related commands and configuration
- iESS: Process buffer is now automatically shown on errors
- ESS: New `ess-switch-to-inferior-or-script-buffer` command bound to `C-c C-z` in both script and process buffers. If invoked form process buffer it switches to the most recent buffer of the same dialect. It is a single key command.
- ESSR-help: On multiple help pages with the same name, `C-c C-v` now asks for user resolution directly in emacs.
- ESS[R] ess-roxy: new variable `ess-roxy-re` for fontification of cases where the number of leading # differs from `ess-roxy-str`.
- ESS[R] Eldoc was considerably enhanced. It now finds hidden default S3 methods and displays non-default methods' arguments after trailing | |.
- ESS[R]: New `ess-display-demos` command bound to `C-c C-d o` and `C-c C-d C-o`
- ESS: New `ess-help-web-search` command bound to `C-c C-d w` and `C-c C-d C-w` to facilitate interactive search of web resources. Implemented for R, Stata and Julia. See also `ess-help-web-search-command`.
- ESS: ess-pdf-viewer-pref accepts now command line arguments
- ESS[Rnw]: Add knitr support. Customize `ess-swv-processor` for the default processor.
- ESS[Rnw]: More thorough renaming of remaining `noweb-*` to `ess-noweb-*`.
- ESS[Rnw] new commands `ess-eval-chunk-and-step` and `ess-eval-chunk` bound to `M-n C-c` and `M-n C-M-x` to mirror standard ess commands in C-c map.
- ESS[R] Auto-completion: new variable `ess-ac-R-argument-suffix` to customize the insertion of trailing "=". Defaults to " = ".
- ESS[Julia]: Added index, apropos and web-search to julia.
- ESS help: More evaluation commands were added to ess-help mode (`C-c C-c`, `C-M-x` etc)

Bug Fixes in 12.09-1:

- iESShelp: Multiple help pages with the same name are properly handled on `C-c C-v`
- iESSremote: Evaluation with ESS remote no longer freezes emacs.
- iESS: `comint-previous-prompt` `C-c C-p` no longer stops on secondary prompt "+".
- iESS[R], iESS(Sqpe) [S] on Windows: The `options("editor")` is now initialized to `emacsclient` instead of the previous `gnuclient`. The user may need to add the line `(server-start)` to the emacs initialization file. `emacsclient` has been included with emacs since GNU Emacs 22.1.
- ESS[Rnw] Fixed "connection to R" bug (in 12.09 only).
- ESS[Rnw] Explicit `ess-swv-stangle` and `ess-swv-sweave` functions.
- ESS[Rnw] Fixed completion and smart underscore problems cause by unmatched "\"'
- ESS[R] is more careful with the R code injection. It now happens only once at the start of the session.
- ESS[R]: Fixed auto-scrolling the comint buffer on evaluation.
- ESS[Julia]: Solve several indentation and word navigation problems.

- ESS[Julia]: Help system works again.

Changes/New Features in 12.09:

- **Due to XEmacs lacking some features that ESS requires, ESS support of XEmacs ends with ESS 12.04-4. This decision will be re-visited in the future as XEmacs continues to sync with GNU Emacs.**

- ESS[R]: On Windows, there is now a new customizable variable (currently called `ess-directory-containing-R`) to tell ESS where to look for the `Rterm.exe` executables. The name of the variable and the values it can take are both in beta and subject to change. Prior to this variable, ESS searched only in the default installation directory. Setting this variable now tells ESS how to find `Rterm.exe` executables when they are installed somewhere else.

- ESS[julia]: *new* mode for editing julia code (`*.jl`). Start with `M-x julia`.

  Full interaction interface, imenu and basic error referencing are available.

- ESS[R] noweb: `noweb-mode` and `noweb-font-lock-mode` have been renamed to `ess-noweb-mode` and `ess-noweb-font-lock-mode` to avoid conflicts with the "real" `noweb-mode`.

- ESS[R] noweb: The long standing font-lock bug has been solved in `ess-noweb` interface.

- ESS: Basic evaluation keys are now bound to `ess-eval-region-*-` functions:
  - `C-M-x` is bound to `ess-eval-region-or-function-or-paragraph`
  - `C-c C-c` is bound to `ess-eval-region-or-function-or-paragraph-and-step`
  - `C-RET` is bound to `ess-eval-region-or-line-and-step`

  Each of these functions first evaluates the region whenever the region is active.

- ESS: `C-M-a`/`C-M-e` now step to beginning/end of paragraph if no function has been detected.

- ESS: `ess-eval-*-and-step` family of functions are now smarter, and don't step to end of buffer or end of chunk code (`@`) when at the end of the code.

- ESS: `ess-handy-commands` function is bound to `C-c h`

- ESS: ESS is now *blinking* the evaluated region. Set `ess-blink-region` to nil to deactivate; `ess-blink-delay` gives the duration of the blink. Evaluated region is "blinked" in `highlight` face.

- ESS[R-help] New key `a` for "apropos()" in help buffers. Also available through `C-c h`.

- ESS[R-help] All R commands of type foo?bar and foo??bar are recognized and redirected into appropriate *ESS-help* buffers.

- ESS[R]: New customization interface for *font-lock*.

  ESS font-lock operates with predefined keywords. Default keywords are listed in `ess-R-font-lock-keywords` and `inferior-R-font-lock-keywords`, which see. The user can easily customize those by adding new keywords. These variables can also be interactively accessed and saved through *ESS/Font-lock* submenu.

  Several new fontification keywords have been added. Most notably the keywords for highlighting of function calls, numbers and operators.

- ESS[R]: auto-complete is now activated by default whenever auto-complete package is detected. Set `ess-use-auto-complete` to nil to deactivate.

- ESS[R]: R AC sources are no longer auto-starting at 0 characters but at the default `ac-auto-start` characters.
- ESS no longer redefines default ac-sources, but only appends `ac-source-filename` to it.
- ESS: `ac-source-R` now concatenates " = " to function arguments.
- ESS: Menus for ESS and iESS have been reorganized and enriched with *Tracebug* and *Developer* submenus.
- ESS[R]: `ess-developer` and `ess-tracebug` commands are available by default in `ess-dev-map` which is bound to *C-c d* in ESS and iESS maps.
- ESS[R]: eldoc truncates long lines whenever `eldoc-echo-area-use-multiline-p` is non-nil (the default). Set this variable to t if you insist on multiline eldoc. See also `ess-eldoc-abbreviation-style`.
- ESS[R]: completion code pre-caches arguments of heavy generics such as `plot` and `print` to eliminated the undesirable delay on first request.
- iESS: Prompts in inferior buffers are now highlighted uniformly with `comint-highlight-prompt` face.
- ESS[R]: R process no longer wait for the completion of input in inferior buffer. Thus, long running commands like `Sys.sleep(5)` no longer stall emacs.
- ESS: [R, S, Stata, Julia] have specialized `ess-X-post-run-hooks`, which are run at the end of subprocess initialization.
- ESS[Stata]: All interactive evaluation commands work as expected. On-line comments are removed before the evaluation and multiline comments are skipped on *C-c C-c* and other interactive commands.
- ESS no longer auto-connects to a subprocess with a different dialect than the current buffer's one.
- ESS: `ess-arg-function-offset-new-line` is now a list for all the ESS indentation styles, which results in the following indentation after an open "(":
  ```
  a <- some.function(other.function(
      arg1,
      arg2)
  ```
- ESS[SAS]: Improved MS RTF support for GNU Emacs; try `ess-sas-rtf-portrait` and `ess-sas-rtf-landscape`.

Changes/Bug Fixes in 12.04-3:
- ESS: basic support for package.el compatibility
- ESS[R]: correct indentation of & and | continuation lines
- `M-x ess-version` shows the svn revision even after `make install`
- ESS[SAS]: improved XEmacs support
- iESS[R]: better finding of previous prompt
- ESS[Stata]: adjusted prompt for mata mode
- ESS[R]: resolved name clashes with cl.el
- ESS[R]: removed dependence on obsolete package assoc

- New `make` target `lisp`, to build the lisp-only part, i.e., not building the docs.

Changes/New Features in 12.04-1:

- iESS[Stata]: New interactive help invocation.
- iESS[Stata]: New custom variable `inferior-STA-start-file`.
- iESS[Stata]: `inferior-STA-program-name` is now "stata" and can be customized.
- ESS[Stata] New sections in stata help files Syntax(s-S), Remarks(r), Title(t).

Bug Fixes in 12.04-1:

- ESS[R]: Better `ess-tracebug` error handling.
- ESS[R]: Corrected `ess-eldoc` help string filtering and improved argument caching.
- ESS[R]: Indentation of non-block if/else/for/while lines fixed.
- `M-x ess-version` should work better.
- ESS: Filename completion now again works inside strings.
- iESS[Stata]: Fixed prompt detection issue.
- ESS[Rd]: R is autostarted also from here, when needed.

Changes/New Features in 12.04:

- ESS: Reverting new behavior of 12.03, `TAB` in `ess-mode` no longer completes by default. If you want smart `TAB` completion in R and S scripts, similarly to iESS behavior, set the variable `ess-tab-complete-in-script` to t. Also see `ess-first-tab-never-complete` for how to customize where first `TAB` is allowed to complete.
- ESS: completion is consistently bound to `M-TAB` (aka `M-C-i`) in both Emacs23 and Emacs24.
- ESS: The variable `ess-arg-function-offset-new-line` introduced in ESS(12.03) now accepts a list with the first element a number to indicate that the offset should be computed from the indent of the previous line. For example setting it to '(2) results in:

```
a <- some.function(
  arg1,
  arg2)
```

Changes/New Features in 12.03:

- ESS indentation: new offset variable `ess-arg-function-offset-new-line` controlling for the indentation of lines immediately following open '('. This is useful to shift backwards function arguments after a long function call expression:

```
a <- some.function(
    arg1,
    arg2)
```

instead of the old

```
a <- some.function(
                arg1,
                arg2)
```

If '(' is not followed by new line the behavior is unchanged:

```
a <- some.function(arg1,
                   arg2)
```

This variable should be set as part of indentation style lists, or in ess-mode hook.

- ESS[R]: *C-c* . sets (indentation) style.

- ESS: In ESS buffers yank(*C-y*) command accepts double argument *C-u C-u* to paste commands only. It deletes any lines not beginning with a prompt, and then removes the prompt from those lines that remain. Useful to paste code from emails, documentation, inferior ESS buffers or transcript files.

- Documentation: ESS user manual has been rearranged and completed with several new chapters and sections to reflect newly added features ("Completion", "Developing with ESS", "ESS tracebug", "ESS developer", "ESS ElDoc", "IDO Completion" and "Evaluating Code")

- RefCard: Reference card was updated to include new features.

- Eldoc: Eldoc was rewritten and is activated by default. See ess-use-eldoc, ess-eldoc-show-on-symbol, ess-eldoc-abbreviation-style variables for how to change the default behavior. *Note:* skeleton-pair-insert-maybe prohibits eldoc display, on (insertion.

- ESS[R]: Eldoc shows arguments of a generic function whenever found.

- ESS: *TAB* in ess-mode now indents and completes, if there is nothing to indent. Set ess-first-tab-never-completes-p to t to make *TAB* never complete on first invocation. Completion mechanism is similar to the completion in the inferior-ess-mode – a filename expansion is tried, if not found ESS completes the symbol by querying the process.

- ESS for emacs version 24 or higher: ESS is fully compatible with the emacs 24 completion scheme, i.e. all the completion is done by completion-at-point. Also in accordance with emacs conventions, ESS doesn't bind *M-TAB* for emacs 24 or higher. *M-TAB* calls the default complete-symbol.

- ESS[R]: Out of the box integration with Auto Completion mode http://cx4a.org/software/auto-complete . Three AC sources ac-source-R-args, ac-source-R-objects and ac-source-R are provided. The last one combines the previous two and makes them play nicely together. Set ess-use-auto-complete to t to start using it. Refer to documentation string of ac-use-auto-complete for further information.

- ESS[R]: New unified and fast argument completion system, comprised of ess-funname.start, ess-function-arguments, ess-get-object-at-point. Eldoc and auto-completion integration are using this system.

- ESS: ess-switch-to-end-of-ESS(*C-c C-z*), and ess-switch-to-ESS(*C-c C-y*): Automatically start the process whenever needed.

- ESS[R]: roxy knows about previewing text version of the documentation. Bound to *C-c C-e t*.

- ESS[R]: Solved the "nil filename" bug in roxygen support.

- ESS[R]: ess-tracebug is now part of ESS:
  New Features:
  - Source injection: Tracebug now can inject source references on the fly during code evaluation, i.e. you don't have to source your file, but just evaluate your code in

normal fashion. Variable `ess-tracebug-inject-source-p` controls this behavior - if t, always inject source reference, if `'function`, inject only for functions (this is the default), if `nil`, never inject.

During the source injection the value of `ess-eval-visibly` is ignored.

- Org-mode support: Visual debugger is now aware of the temporary org source editing buffer (`C-c '`) and jumps through this buffers if still alive, or in original org buffer otherwise.

- New keys in watch mode: *?* and *d*

- Two new hooks: ess-tracebug-enter-hook and ess-tracebug-exit-hook

- ESS[R]: New package `ess-developer` to evaluate R code directly in the package environment and namespace. It can be toggled on and off with *C-c d t*. When `ess-developer` is on all ESS evaluation commands are redefined to evaluate code in appropriate environments. Add package names to the list of your development packages with *C-d a*, and remove with *C-d r*. Source the current file with *C-d s*.Evaluation function which depend on `'ess-eval-region'` ask for the package to source the code into, `ess-eval-function` and alternatives search for the function name in the development packages' environment and namespace and insert the definition accordingly. See the documentation section "Developing with ESS/ESS developer" for more details.

- ESS[R] help system:

New Features:

- *q* quits window instead of calling `ess-switch-to-end-of-ESS`. This is consistent with emacs behavior help and other special buffers (*breaking change*).

- *k* kills window without asking for the name (pointed by Sam Steingold)

- Help map inherits from `special-mode-map` as sugested by Sam Steingold.

- Package index: new function `ess-display-index` bound to *i* in help mode map.

- Package vignettes: new function `ess-display-vignettes` bound to *v* in help mode map.

- Display help in HTML browser: new function `ess-display-help-in-browser` bound to *w* in help mode map. It depends on R's `browser` option.

- New custom variable `ess-help-pop-to-buffer`: if non-nil ESS help buffers are given focus on display. The default is t (*breaking change*).

- New menu entries for the above functions.

- Bogus help buffers are no longer generated by default, i.e. buffers of the form "No documentation for 'foo' in specified packages and libraries: you could try '??foo' ". `ess-help-kill-bogus-buffers` now defaults to t. Beware, there may be instances where the default is unsatisfactory such as debugging and/or during R development. Thanks to Ross Boylan for making the suggestion, Sam Steingold for reminding us of this variable and Martin Maechler for the warning.

- ESS now uses `IDO` completing read functionality for all the interactive requests. It uses ido completion mechanism whenever available, and falls back on classical completing-read otherwise. You can set `ess-use-ido` to nil if you don't want the IDO completion. See the documentation string of `ess-use-ido` for more information about `IDO` and ESS configuration.

- ESS[S]: "," " is bound to ess-smart-comma: If comma is invoked at the process marker of an ESS inferior buffer, request and execute a command from 'ess-handy-commands' list. If ess-R-smart-operators is t 'ess-smart-comma also inserts " " after comma.

- ESS[S], notably R: Variable 'ess-handy-commands' stores an alist of useful commands which are called by ess-smart-comma in the inferior buffer.

  Currently containing:

  change-directory
  : `ess-change-directory`

  help-index `ess-display-index`

  help-object
  : `ess-display-help-on-object`

  vignettes `ess-display-vignettes`

  objects[ls] `ess-execute-objects`

  search `ess-execute-search`

  set-width `ess-execute-screen-options`

  install.packages
  : `ess-install.packages`

  library `ess-library`

  setRepos `ess-setRepositories`

  sos `ess-sos`

  Handy commands: `ess-library`, `ess-install.packages`, etc - ask for item with completion and execute the correspond command. `ess-sos` is a interface to `findFn` function in package `sos`. If package `sos` is not found, ask user for interactive install.

- ESS: New dynamic mode line indicator: Process status is automatically reflected in all mode-lines of associated with the process buffers. Particularly useful for displaying debug status of `ess-tracebug` and developer status of `ess-developer` in all associated buffers.

- ESS: New `ess-completing-read` mechanism: ESS uses `ido` completions whenever possible. Variable `ess-use-ido` controls whether to use ido completion or not. Active by default.

- ESS now supports comint fields for output and input detection. This feature is not used by default, but might be useful in the future.

- ESS[S]: New custom variable `inferior-ess-S-prompt` to customize prompt detection regular expression in the inferior ESS buffers. You can customize this variable to enhance comint navigation (`comint-previous-prompt` and `comint-next-prompt`) the inferior buffers.

- ESS[R]: Internal R completion retrieval (`ess-R-complete-object-name`) was rewritten and is faster now.

- ESS is using process plist to store process specific variables, as opposed to buffer local variables as it was using before. The use of buffer local variables to store process variables is discouraged.

- ESS: new functions to manipulate process plists: `ess-process-get` and `ess-process-set`.

- ESS: Internal process waiting mechanism was completely rewritten. ESS no more relies on prompt regular expressions for the prompt detection. The only requirement on the primary process prompt is to end in `>`. This could be overwritten by setting `inferor-ess-primary-prompt`.

- ESS[S], notably `R`: Saved command history: *ess-history-file* now accepts `t` (default), `nil`, or a file name. By setting it to `nil` no command line history is saved anymore. *ess-history-directory* now allows to have the history all saved in one "central" file.

- ESS[R]: more Roxygen improvements.

- ESS[R]: `C-c .` to set (indentation) style.

- ESS[R]: Functions with non-standard names (for example 'aaa-bbb:cc') are properly handled by font-lock and evaluation routines.

- ESS[R]:Several regexp bugs (described in etc/R-ESS-bugs.el) were fixed in `ess-get-words-from-vector` and `ess-command`.

1.3 Authors of and contributors to ESS

The ESS environment is built on the open-source projects of many contributors, dating back to 1989 where Doug Bates and Ed Kademan wrote S-mode to edit S and Splus files in GNU Emacs. Frank Ritter and Mike Meyer added features, creating version 2. Meyer and David Smith made further contributions, creating version 3. For version 4, David Smith provided significant enhancements to allow for powerful process interaction.

John Sall wrote GNU Emacs macros for SAS source code around 1990. Tom Cook added functions to submit jobs, review listing and log files, and produce basic views of a dataset, thus creating a SAS-mode which was distributed in 1994.

In 1994, A.J. Rossini extended S-mode to support XEmacs. Together with extensions written by Martin Maechler, this became version 4.7 and supported S, Splus, and R. In 1995, Rossini extended SAS-mode to work with XEmacs.

In 1997, Rossini merged S-mode and SAS-mode into a single Emacs package for statistical programming; the product of this marriage was called ESS version 5. Richard M. Heiberger designed the inferior mode for interactive SAS and SAS-mode was further integrated into ESS. Thomas Lumley's Stata mode, written around 1996, was also folded into ESS. More changes were made to support additional statistical languages, particularly XLispStat.

ESS initially worked only with Unix statistics packages that used standard-input and standard-output for both the command-line interface and batch processing. ESS could not communicate with statistical packages that did not use this protocol. This changed in 1998 when Brian Ripley demonstrated use of the Windows Dynamic Data Exchange (DDE) protocol with ESS. Heiberger then used DDE to provide interactive interfaces for Windows versions of Splus. In 1999, Rodney A. Sparapani and Heiberger implemented SAS batch for ESS relying on files, rather than standard-input/standard-output, for Unix, Windows and Mac. In 2001, Sparapani added BUGS batch file processing to ESS for Unix and Windows.

- The multiple process code, and the idea for `ess-eval-line-and-next-line` are by Rod Ball.

- Thanks to Doug Bates for many useful suggestions.

- Thanks to Martin Maechler for reporting and fixing bugs, providing many useful comments and suggestions, and for maintaining the ESS mailing lists.
- Thanks to Frank Ritter for updates, particularly the menu code, and invaluable comments on the manual.
- Thanks to Ken'ichi Shibayama for his excellent indenting code, and many comments and suggestions.
- Thanks to Aki Vehtari for adding interactive BUGS support.
- Thanks to Brendan Halpin for bug-fixes and updates to Stata-mode.
- Last, but definitely not least, thanks to the many ESS users and contributors to the ESS mailing lists.

ESS is being developed and currently maintained by

- A.J. Rossini
- Richard M. Heiberger
- Kurt Hornik
- Martin Maechler
- Rodney A. Sparapani
- Stephen Eglen
- Sebastian P. Luque
- Henning Redestig
- Vitalie Spinu

1.4 Getting the latest version of ESS

The latest released version of ESS is always available on the web at: ESS web page or StatLib

1.4.1 Git for ESS development

For development and experimentation on new ESS features, there is now a GitHub branch for ESS, available at `https://github.com/emacs-ess/ESS`.

1.5 How to read this manual

If you need to install ESS, read Chapter 2 [Installation], page 20 for details on what needs to be done before proceeding to the next chapter.

In this manual we use the standard notation for describing the keystrokes used to invoke certain commands. `C-<chr>` means hold the CONTROL key while typing the character <chr>. `M-<chr>` means hold the META or EDIT or ALT key down while typing <chr>. If there is no META, EDIT or ALT key, instead press and release the ESC key and then type <chr>.

All ESS commands can be invoked by typing `M-x command`. Most of the useful commands are bound to keystrokes for ease of use. Also, the most popular commands are also available through the emacs menubar, and finally, if available, a small subset are provided on the toolbar. Where possible, keybindings are similar to other modes in emacs to strive for a consistent user interface within emacs, regardless of the details of which programming language is being edited, or process being run.

Some commands, such as *M-x R* can accept an optional 'prefix' argument. To specify the prefix argument, you would type *C-u* before giving the command. e.g. If you type *C-u M-x R*, you will be asked for command line options that you wish to invoke the R process with.

Emacs is often referred to as a 'self-documenting' text editor. This applies to ESS in two ways. First, limited documentation about each ESS command can be obtained by typing *C-h f*. For example, if you type *C-h f ess-eval-region*, documentation for that command will appear in a separate *Help* buffer. Second, a complete list of keybindings that are available in each ESS mode and brief description of that mode is available by typing *C-h m* within an ESS buffer.

Emacs is a versatile editor written in both C and lisp; ESS is written in the Emacs lisp dialect (termed 'elisp') and thus benefits from the flexible nature of lisp. In particular, many aspects of ESS behaviour can be changed by suitable customization of lisp variables. This manual mentions some of the most frequent variables. A full list of them however is available by using the Custom facility within emacs. (Type *M-x customize-group RET ess RET* to get started.) Appendix A [Customization], page 82 provides details of common user variables you can change to customize ESS to your taste, but it is recommended that you defer this section until you are more familiar with ESS.

2 Installing ESS on your system

The following section details those steps necessary to get ESS running on your system.

2.1 Step by step instructions

1. Download the latest zip or tgz archive from ESS downloads area and unpack it into a directory where you would like ESS to reside. We will denote this directory as /path/to/ESS/ hereafter.

   Alternatively you can use git to fetch the most recent development version to your local machine with:

   ```
   git clone https://github.com/emacs-ess/ESS.git /path/to/ESS
   ```

2. **Optionally**, compile elisp files and build the documentation with:

   ```
   cd /path/to/ESS/
   make
   ```

   Without this step, info, pdf and html documentation and reference card will not be available.

3. **Optionally**, install into your local machine with make install. You might need administrative privileges:

   ```
   make install
   ```

   The files are installed into /usr/share/emacs directory. For this step to run correctly on Mac OS X, you will need to adjust the PREFIX path in Makeconf. The necessary code and instructions are commented in that file.

4. If you have performed the make install step from above, just add

   ```
   (require 'ess-site)
   ```

   to your ~/.emacs file. Otherwise, you should add /path/to/ESS/lisp/ to your emacs load path and then load ESS with the following lines in your ~/.emacs:

   ```
   (add-to-list 'load-path "/path/to/ESS/lisp/")
   (load "ess-site")
   ```

5. Restart your Emacs and check that ESS was loaded from a correct location with M-x ess-version.

Note for Windows and Mac OS X users: The most straightforward way to install Emacs on your machine is by downloading all-in-one Emacs binary by Vincent Goulet.

Note for XEmacs users: Due to XEmacs lacking some features that ESS requires, ESS support of XEmacs ends with ESS 12.04-4. This decision will be re-visited in the future as XEmacs continues to sync with GNU Emacs.

2.2 License

The source and documentation of ESS is free software. You can redistribute it and/or modify it under the terms of the GNU General Public License as published by the Free Software Foundation; either version 2, or (at your option) any later version.

ESS is distributed in the hope that it will be useful, but WITHOUT ANY WARRANTY; without even the implied warranty of MERCHANTABILITY or FITNESS FOR A PARTICULAR PURPOSE. See the GNU General Public License in the file COPYING in the same directory as this file for more details.

2.3 Stability

All recent released versions are meant to be release-quality versions. While some new features are being introduced, we are cleaning up and improving the interface. We know that there are many remaining opportunities for documentation improvements, but all contributors are volunteers and time is precious. Patches or suggested fixes with bug reports are much appreciated!

2.4 Requirements

ESS is most likely to work with current/recent versions of the following statistical packages: R/S-PLUS, SAS, Stata, OpenBUGS and JAGS.

ESS supports current, and recent, stable versions of GNU Emacs (currently, specifically, the 23.x and 24.x series; alpha/beta/pre-release versions are NOT SUPPORTED). Non-Windows users beware: GNU Emacs 24.3 is preferable to 24.1 or 24.2: these broken builds suffer from bug 12463 http://debbugs.gnu.org/cgi/bugreport.cgi?bug=12463 which will cause emacs and ESS to get progressively slower over time.

Due to XEmacs lacking some features that ESS requires, ESS support of XEmacs ends with ESS 12.04-4. This decision will be re-visited in the future as XEmacs continues to sync with GNU Emacs.

To build the PDF documentation, you will need a version of TeX Live or texinfo that includes texi2dvi (BEWARE: recent TeX Live, and some texinfo RPMs, do NOT include texi2dvi).

3 Interacting with statistical programs

As well as using ESS to edit your source files for statistical programs, you can use ESS to run these statistical programs. In this chapter, we mostly will refer by example to running S from within emacs. The emacs convention is to name such processes running under its control as 'inferior processes'. This term can be slightly misleading, in which case these processes can be thought of 'interactive processes'. Either way, we use the term 'iESS' to refer to the Emacs mode used to interact with statistical programs.

3.1 Starting an ESS process

To start an S session on Unix or on Windows when you use the Cygwin bash shell, simply type *M-x S RET*.

To start an S session on Windows when you use the MSDOS prompt shell, simply type *M-x S+6-msdos RET*.

S will then (by default) ask the question

    S starting data directory?

Enter the name of the directory you wish to start S from (that is, the directory you would have cd'd to before starting S from the shell). This directory should have a .Data subdirectory.

You will then be popped into a buffer with name '*S*' which will be used for interacting with the ESS process, and you can start entering commands.

3.2 Running more than one ESS process

ESS allows you to run more than one ESS process simultaneously in the same session. Each process has a name and a number; the initial process (process 1) is simply named (using S-PLUS as an example) 'S+3:1'. The name of the process is shown in the mode line in square brackets (for example, '[S+3:2]'); this is useful if the process buffer is renamed. Without a prefix argument, *M-x S* starts a new ESS process, using the first available process number. With a prefix argument (for R), *C-u M-x R* allows for the specification of command line options.

You can switch to any active ESS process with the command 'M-x ess-request-a-process'. Just enter the name of the process you require; completion is provided over the names of all running S processes. This is a good command to consider binding to a global key.

3.3 ESS processes on Remote Computers

ESS works with processes on remote computers as easily as with processes on the local machine. The recommended way to access a statistical program on remote computer is to start it with tramp. Require tramp in your .emacs file:

    (require 'tramp)

Now start an ssh session with 'C-x f /ssh:user@host: RET'. Tramp should open a dired buffer in your remote home directory. Now call your favorite ESS process (R, Julia, stata etc) as you would usually do on local machine: M-x R.

Alternatively you can start your process normally (M-x R). After you are asked for starting directory, simply type '/ssh:user@host: RET'. R process will be started on the remote machine.

To simplify the process even further create a "config" file in your .ssh/ folder and add an account. For example if you use amazon EC2, it might look like following:

```
Host amazon
    Hostname ec2-54-215-203-181.us-west-1.compute.amazonaws.com
    User ubuntu
    IdentityFile ~/.ssh/my_amazon_key.pem
    ForwardX11 yes
```

With this configuration /ssh:amazon: is enough to start a connection. The ForwardX11 is needed if you want to see R graphic device showing on the current machine

Other ways to setup a remote ESS connection are through ess-remote.

1. Start a new shell, telnet or ssh buffer and connect to the remote computer (e.g. use, 'M-x shell', 'M-x telnet' or 'M-x ssh'; ssh.el is available at http://www.splode.com/~friedman/software/emacs-lisp/src/ssh.el).

2. Start the ESS process on the remote machine, for example with one of the commands 'Splus', or 'R', or 'sas -stdio'.

3. Start 'M-x ess-remote'. You will be prompted for a program name with completion. Choose one. Your process is now known to ESS. All the usual ESS commands ('C-c C-n' and its relatives) now work with the S language processes. For SAS you need to use a different command 'C-c i' (that is a regular 'i', not a 'C-i') to send lines from your myfile.sas to the remote SAS process. 'C-c i' sends lines over invisibly. With ess-remote you get teletype behavior—the data input, the log, and the listing all appear in the same buffer. To make this work, you need to end every PROC and DATA step with a "RUN;" statement. The "RUN;" statement is what tells SAS that it should process the preceding input statements.

4. Graphics (interactive) on the remote machine. If you run X11 (See Section 11.5.2 [X11], page 58, X Windows) on both the local and remote machines then you should be able to display the graphs locally by setting the 'DISPLAY' environment variable appropriately. Windows users can download 'xfree86' from cygwin.

5. Graphics (static) on the remote machine. If you don't run the X window system on the local machine, then you can write graphics to a file on the remote machine, and display the file in a graphics viewer on the local machine. Most statistical software can write one or more of postscript, GIF, or JPEG files. Depending on the versions of emacs and the operating system that you are running, emacs itself may display '.gif' and '.jpg' files. Otherwise, a graphics file viewer will be needed. Ghostscript/ghostview may be downloaded to display '.ps' and '.eps' files. Viewers for GIF and JPEG are usually included with operating systems. See Section 13.5 [ESS(SAS)–Function keys for batch processing], page 71, for more information on using the F12 key for displaying graphics files with SAS.

Should you or a colleague inadvertently start a statistical process in an ordinary '*shell*' buffer, the 'ess-remote' command can be used to convert it to an ESS buffer and allow you to use the ESS commands with it.

We have two older commands, now deprecated, for accessing ESS processes on remote computers. See Section 3.4 [S+elsewhere and ESS-elsewhere], page 24.

3.4 S+elsewhere and ESS-elsewhere

These commands are now deprecated. We recommend 'ess-remote'. We have two versions of the elsewhere function. 'S+elsewhere' is specific for the S-Plus program. The more general function 'ESS-elsewhere' is not as stable.

1. Enter 'M-x S+elsewhere'. You will be prompted for a starting directory. I usually give it my project directory on the local machine, say '~myname/myproject/'

   Or enter 'M-x ESS-elsewhere'. You will be prompted for an ESS program and for a starting directory. I usually give it my project directory on the local machine, say '~myname/myproject/'

2. The '*S+3*' buffer will appear with a prompt from the local operating system (the unix prompt on a unix workstation or with cygwin bash on a PC, or the msdos prompt on a PC without bash). emacs may freeze because the cursor is at the wrong place. Unfreeze it with 'C-g' then move the cursor to the end with 'M->'. With 'S+elsewhere' the buffer name is based on the name of the ESS program.

3. Enter 'telnet myname@other.machine' (or 'ssh myname@other.machine'). You will be prompted for your password on the remote machine. Use 'M-x send-invisible' before typing the password itself.

4. Before starting the ESS process, type 'stty -echo nl' at the unix prompt. The '-echo' turns off the echo, the 'nl' turns off the newline that you see as '^M'.

5. You are now talking to the unix prompt on the other machine in the '*S+3*' buffer. cd into the directory for the current project and start the ESS process by entering 'Splus' or 'R' or 'sas -stdio' as appropriate. If you can login remotely to your Windows 2000, then you should be able to run 'Sqpe' on the Windows machine. I haven't tested this and no-one has reported their tests to me. You will not be able to run the GUI through this text-only connection.

6. Once you get the S or R or SAS prompt, then you are completely connected. All the 'C-c C-n' and related commands work correctly in sending commands from 'myfile.s' or 'myfile.r' on the PC to the '*S+3*' buffer running the S or R or SAS program on the remote machine.

7. Graphics on the remote machine works fine. If you run the X window system on the remote unix machine you should be able to display them in 'xfree86' on your PC. If you don't run X Windows, or X11, then you can write graphics to the postscript device and copy it to your PC with dired and display it with ghostscript.

3.5 Changing the startup actions

If you do not wish ESS to prompt for a starting directory when starting a new process, set the variable ess-ask-for-ess-directory to nil. In this case, the starting directory will be set using one of the following methods:

1. If the variable ess-directory-function stores the name of a function, the value returned by this function is used. The default for this variable is nil.

2. Otherwise, if the variable `ess-directory` stores the name of a directory (ending in a slash), this value is used. The default for this variable is nil.

3. Otherwise, the working directory of the current buffer is used.

If `ess-ask-for-ess-directory` has a non-`nil` value (as it does by default) then the value determined by the above rules provides the default when prompting for the starting directory. Incidentally, `ess-directory` is an ideal variable to set in `ess-pre-run-hook`.

If you like to keep a record of your S sessions, set the variable `ess-ask-about-transfile` to `t`, and you will be asked for a filename for the transcript before the ESS process starts.

`ess-ask-about-transfile`   [User Option]
   If non-`nil`, as for a file name in which to save the session transcript.

Enter the name of a file in which to save the transcript at the prompt. If the file doesn't exist it will be created (and you should give it a file name ending in '.St'); if the file already exists the transcript will be appended to the file. (Note: if you don't set this variable but you still want to save the transcript, you can still do it later — see Section 4.2.4 [Saving transcripts], page 28.)

Once these questions are answered (if they are asked at all) the S process itself is started by calling the program name specified in the variable `inferior-ess-program`. If you need to pass any arguments to this program, they may be specified in the variable `inferior-S_program_name-args` (e.g. if `inferior-ess-program` is `"S+"` then the variable to set is `inferior-S+-args`. It is not normally necessary to pass arguments to the S program; in particular do not pass the '-e' option to `Splus`, since ESS provides its own command history mechanism.

By default, the new process will be displayed in the same window in the current frame. If you wish your S process to appear in a separate variable, customize the variable `inferior-ess-own-frame`. Alternatively, change `inferior-ess-same-window` if you wish the process to appear within another window of the current frame.

4 Interacting with the ESS process

The primary function of the ESS package is to provide an easy-to-use front end to the S interpreter. This is achieved by running the S process from within an Emacs buffer, called hereafter *inferior* buffer, which has an active `inferior-ess-mode`. The features of Inferior S mode are similar to those provided by the standard Emacs shell mode (see Section "Shell Mode" in *The Gnu Emacs Reference Manual*). Command-line completion of S objects and a number of 'hot keys' for commonly-used S commands are also provided for ease of typing.

4.1 Entering commands and fixing mistakes

Sending a command to the ESS process is as simple as typing it in and pressing the RETURN key:

`inferior-ess-send-input`   [Command]

    RET Send the command on the current line to the ESS process.

If you make a typing error before pressing *RET* all the usual Emacs editing commands are available to correct it (see Section "Basic editing commands" in *The GNU Emacs Reference Manual*). Once the command has been corrected you can press RETURN (even if the cursor is not at the end of the line) to send the corrected command to the ESS process.

Emacs provides some other commands which are useful for fixing mistakes:

C-c C-w   `backward-kill-word` Deletes the previous word (such as an object name) on the command line.

C-c C-u   `comint-kill-input` Deletes everything from the prompt to point. Use this to abandon a command you have not yet sent to the ESS process.

C-c C-a   `comint-bol` Move to the beginning of the line, and then skip forwards past the prompt, if any.

See Section "Shell Mode" in *The Gnu Emacs Reference Manual*, for other commands relevant to entering input.

4.2 Manipulating the transcript

Most of the time, the cursor spends most of its time at the bottom of the ESS process buffer, entering commands. However all the input and output from the current (and previous) ESS sessions is stored in the process buffer (we call this the transcript) and often we want to move back up through the buffer, to look at the output from previous commands for example.

Within the process buffer, a paragraph is defined as the prompt, the command after the prompt, and the output from the command. Thus *M-{* and *M-}* move you backwards and forwards, respectively, through commands in the transcript. A particularly useful command is *M-h* (`mark-paragraph`) which will allow you to mark a command and its entire output (for deletion, perhaps). For more information about paragraph commands, see Section "Paragraphs" in *The GNU Emacs Reference Manual*.

If an ESS process finishes and you restart it in the same process buffer, the output from the new ESS process appears after the output from the first ESS process separated by a form-feed ('^L') character. Thus pages in the ESS process buffer correspond to ESS sessions.

Thus, for example, you may use `C-x [` and `C-x ]` to move backward and forwards through ESS sessions in a single ESS process buffer. For more information about page commands, see Section "Pages" in *The GNU Emacs Reference Manual*.

4.2.1 Manipulating the output from the last command

Viewing the output of the command you have just entered is a common occurrence and ESS provides a number of facilities for doing this. Whenever a command produces a longish output, it is possible that the window will scroll, leaving the next prompt near the middle of the window. The first part of the command output may have scrolled off the top of the window, even though the entire output would fit in the window if the prompt were near the bottom of the window. If this happens, you can use the following comint commands:

`comint-show-maximum-output` to move to the end of the buffer, and place cursor on bottom line of window to make more of the last output visible. To make this happen automatically for all inputs, set the variable `comint-scroll-to-bottom-on-input` to `t` or `'this`. If the first part of the output is still not visible, use `C-c C-r` (`comint-show-output`), which moves cursor to the previous command line and places it at the top of the window.

Finally, if you want to discard the last command output altogether, use `C-c C-o` (`comint-kill-output`), which deletes everything from the last command to the current prompt. Use this command judiciously to keep your transcript to a more manageable size.

4.2.2 Viewing older commands

If you want to view the output from more historic commands than the previous command, commands are also provided to move backwards and forwards through previously entered commands in the process buffer:

`C-c C-p`   `comint-previous-input` Moves point to the preceding command in the process buffer.

`C-c C-n`   `comint-next-input` Moves point to the next command in the process buffer.

Note that these two commands are analogous to `C-p` and `C-n` but apply to command lines rather than text lines. And just like `C-p` and `C-n`, passing a prefix argument to these commands means to move to the *ARG*'th next (or previous) command. (These commands are also discussed in Section "Shell History Copying" in *The GNU Emacs Reference Manual*.)

There are also two similar commands (not bound to any keys by default) which move to preceding or succeeding commands, but which first prompt for a regular expression (see Section "Syntax of Regular Expression" in *The GNU Emacs Reference Manual*), and then moves to the next (previous) command matching the pattern.

`comint-backward-matching-input regexp arg`
`comint-forward-matching-input regexp arg`

Search backward (forward) through the transcript buffer for the *arg*'th previous (next) command matching *regexp*. *arg* is the prefix argument; *regexp* is prompted for in the minibuffer.

4.2.3 Re-submitting commands from the transcript

When moving through the transcript, you may wish to re-execute some of the commands you find there. ESS provides three commands to do this; these commands may be used whenever the cursor is within a command line in the transcript (if the cursor is within some command *output*, an error is signalled). Note all three commands involve the RETURN key.

RET   inferior-ess-send-input See Section 4.1 [Command-line editing], page 26.

C-c RET   comint-copy-old-input Copy the command under the cursor to the current command line, but don't execute it. Leaves the cursor on the command line so that the copied command may be edited.

When the cursor is not after the current prompt, the RETURN key has a slightly different behavior than usual. Pressing RET on any line containing a command that you entered (i.e. a line beginning with a prompt) sends that command to the ESS process once again. If you wish to edit the command before executing it, use C-c RET instead; it copies the command to the current prompt but does not execute it, allowing you to edit it before submitting it.

These commands work even if the current line is a continuation line (i.e. the prompt is '+' instead of '>') — in this case all the lines that form the multi-line command are concatenated together and the resulting command is sent to the ESS process (currently this is the only way to resubmit a multi-line command to the ESS process in one go). If the current line does not begin with a prompt, an error is signalled. This feature, coupled with the command-based motion commands described above, could be used as a primitive history mechanism. ESS provides a more sophisticated mechanism, however, which is described in Section 4.3 [Command History], page 28.

4.2.4 Keeping a record of your S session

To keep a record of your S session in a disk file, use the Emacs command C-x C-w (write-file) to attach a file to the ESS process buffer. The name of the process buffer will (probably) change to the name of the file, but this is not a problem. You can still use S as usual; just remember to save the file before you quit Emacs with C-x C-s. You can make ESS prompt you for a filename in which to save the transcript every time you start S by setting the variable ess-ask-about-transfile to t; See Section 3.5 [Customizing startup], page 24. We recommend you save your transcripts with filenames that end in '.St'. There is a special mode (ESS transcript mode — see Chapter 6 [Transcript Mode], page 36) for editing transcript files which is automatically selected for files with this suffix.

S transcripts can get very large, so some judicious editing is appropriate if you are saving it in a file. Use C-c C-o whenever a command produces excessively long output (printing large arrays, for example). Delete erroneous commands (and the resulting error messages or other output) by moving to the command (or its output) and typing M-h C-w. Also, remember that C-c C-x (and other hot keys) may be used for commands whose output you do not wish to appear in the transcript. These suggestions are appropriate even if you are not saving your transcript to disk, since the larger the transcript, the more memory your Emacs process will use on the host machine.

Finally, if you intend to produce S source code (suitable for using with source() or inclusion in an S function) from a transcript, then the command ess-transcript-clean-region may be of use. see Section 6.2 [Clean], page 36

4.3 Command History

ESS provides easy-to-use facilities for re-executing or editing previous commands. An input history of the last few commands is maintained (by default the last 500 commands are stored, although this can be changed by setting the variable `comint-input-ring-size` in `inferior-ess-mode-hook`.) The simplest history commands simply select the next and previous commands in the input history:

M-p   `comint-previous-input` Select the previous command in the input history.

M-n   `comint-next-input` Select the next command in the input history.

For example, pressing *M-p* once will re-enter the last command into the process buffer after the prompt but does not send it to the ESS process, thus allowing editing or correction of the command before the ESS process sees it. Once corrections have been made, press *RET* to send the edited command to the ESS process.

If you want to select a particular command from the history by matching it against a regular expression (see Section "Syntax of Regular Expression" in *The GNU Emacs Reference Manual*), to search for a particular variable name for example, these commands are also available:

M-r   `comint-history-isearch-backward-regexp` Prompt for a regular expression, and search backwards through the input history for a command matching the expression.

A common type of search is to find the last command that began with a particular sequence of characters; the following two commands provide an easy way to do this:

C-c M-r   `comint-previous-matching-input-from-input` Select the previous command in the history which matches the string typed so far.

C-c M-s   `comint-next-matching-input-from-input` Select the next command in the history which matches the string typed so far.

Instead of prompting for a regular expression to match against, as they instead select commands starting with those characters already entered. For instance, if you wanted to re-execute the last `attach()` command, you may only need to type `att` and then *C-c M-r* and *RET*.

See Section "Shell History Ring" in *The GNU Emacs Reference Manual*, for a more detailed discussion of the history mechanism, and do experiment with the `In/Out` menu to explore the possibilities.

Many ESS users like to have even easier access to these, and recommend to add something like

```
(eval-after-load "comint"
 '(progn
    (define-key comint-mode-map [up]
      'comint-previous-matching-input-from-input)
    (define-key comint-mode-map [down]
      'comint-next-matching-input-from-input)

    ;; also recommended for ESS use --
    (setq comint-scroll-to-bottom-on-output 'others)
    (setq comint-scroll-show-maximum-output t)
    ;; somewhat extreme, almost disabling writing in *R*, *shell* buffers above
```

```
            (setq comint-scroll-to-bottom-on-input 'this)
            ))
```

to your `.emacs` file, where the last two settings are typically desirable for the situation where you work with a script (e.g., `code.R`) and send code chunks to the process buffer (e.g. `*R*`). Note however that these settings influence all `comint`-using emacs modes, not just the ESS ones, and for that reason, these customization cannot be part of ESS itself.

4.3.1 Saving the command history

The `ess-history-file` variable, which is `t` by default, together with `ess-history-directory`, governs if and where the command history is saved and restored between sessions. By default, `ess-history-directory` is `nil`, and the command history will be stored (as text file) in the `ess-directory`, e.g., as `.Rhistory`.

Experienced ESS users often work exclusively with script files rather than in a (e.g., `*R`) console session, and may not want to save any history files, and hence have:

```
        (setq ess-history-file nil)
```

or will only want one global command history file and have:

```
        (setq ess-history-directory "~/.R/")
```

in your `.emacs` file.

4.4 References to historical commands

Instead of searching through the command history using the command described in the previous section, you can alternatively refer to a historical command directly using a notation very similar to that used in `csh`. History references are introduced by a '!' or '^' character and have meanings as follows:

'`!!`'   The immediately previous command

'`!-N`'   The Nth previous command

'`!text`'   The last command beginning with the string '`text`'

'`!?text`'   The last command containing the string '`text`'

In addition, you may follow the reference with a *word designator* to select particular *words* of the input. A word is defined as a sequence of characters separated by whitespace. (You can modify this definition by setting the value of `comint-delimiter-argument-list` to a list of characters that are allowed to separate words and themselves form words.) Words are numbered beginning with zero. The word designator usually begins with a '`:`' (colon) character; however it may be omitted if the word reference begins with a '^', '$', '*' or '-'. If the word is to be selected from the previous command, the second '!' character can be omitted from the event specification. For instance, '`!!:1`' and '`!:1`' both refer to the first word of the previous command, while '`!!$`' and '`!$`' both refer to the last word in the previous command. The format of word designators is as follows:

'`0`'   The zeroth word (i.e. the first one on the command line)

'`n`'   The nth word, where n is a number

'`^`'   The first word (i.e. the second one on the command line)

'$'   The last word

'x-y'   A range of words; '-y' abbreviates '0-y'

'*'   All the words except the zeroth word, or nothing if the command had just one
              word (the zeroth)

'x*'   Abbreviates x-$

'x-'   Like 'x*', but omitting the last word

In addition, you may surround the entire reference except for the first '!' by braces to allow it to be followed by other (non-whitespace) characters (which will be appended to the expanded reference).

Finally, ESS also provides quick substitution; a reference like '^old^new^' means "the last command, but with the first occurrence of the string 'old' replaced with the string 'new'" (the last '^' is optional). Similarly, '^old^' means "the last command, with the first occurrence of the string 'old' deleted" (again, the last '^' is optional).

To convert a history reference as described above to an input suitable for S, you need to *expand* the history reference, using the TAB key. For this to work, the cursor must be preceded by a space (otherwise it would try to complete an object name) and not be within a string (otherwise it would try to complete a filename). So to expand the history reference, type *SPC TAB*. This will convert the history reference into an S command from the history, which you can then edit or press RET to execute.

For example, to execute the last command that referenced the variable `data`, type *!?data SPC TAB RET*.

4.5 Hot keys for common commands

ESS provides a number of commands for executing the commonly used functions. These commands below are basically information-gaining commands (such as `objects()` or `search()`) which tend to clutter up your transcript and for this reason some of the hot keys display their output in a temporary buffer instead of the process buffer by default. This behavior is controlled by the following option:

ess-execute-in-process-buffer   [User Option]
       If non-`nil`, means that these commands will produce their output in the process
       buffer instead.

In any case, passing a prefix argument to the commands (with *C-u*) will reverse the meaning of **ess-execute-in-process-buffer** for that command, i.e. the output will be displayed in the process buffer if it usually goes to a temporary buffer, and vice-versa. These are the hot keys that behave in this way:

ess-execute-objects *posn*   [Command]
       C-c C-x Sends the `objects()` command to the ESS process. A prefix argument spec-
       ifies the position on the search list (use a negative argument to toggle **ess-execute-
       in-process-buffer** as well). A quick way to see what objects are in your working
       directory. A prefix argument of 2 or more means get objects for that position. A
       negative prefix argument *posn* gets the objects for that position, as well as toggling
       ess-execute-in-process-buffer.

ess-execute-search *invert*   [Command]

> *C-c C-s* Sends the **inferior-ess-search-list-command** command to the **ess-language** process; **search()** in S. Prefix *invert* toggles **ess-execute-in-process-buffer**.

ess-execute may seem pointless when you could just type the command in anyway, but it proves useful for 'spot' calculations which would otherwise clutter your transcript, or for evaluating an expression while partway through entering a command. You can also use this command to generate new hot keys using the Emacs keyboard macro facilities; see Section "Keyboard Macros" in *The GNU Emacs Reference Manual*.

The following hot keys do not use **ess-execute-in-process-buffer** to decide where to display the output — they either always display in the process buffer or in a separate buffer, as indicated:

ess-load-file *filename*   [Command]

> *C-c C-l* Prompts for a file (*filename*) to load into the ESS process using **source()**. If there is an error during loading, you can jump to the error in the file with the following function.

ess-parse-errors *arg reset*   [Command]

> *C-c '* or *C-x '* Visits next **next-error** message and corresponding source code. If all the error messages parsed so far have been processed already, the message buffer is checked for new ones. A prefix *arg* specifies how many error messages to move; negative means move back to previous error messages. Just *C-u* as a prefix means reparse the error message buffer and start at the first error. The *reset* argument specifies restarting from the beginning.
>
> See Section 7.3 [Error Checking], page 38, for more details.

ess-display-help-on-object *object command*   [Command]

> *C-c C-v* Pops up a help buffer for an S *object* or function. If *command* is supplied, it is used instead of **inferior-ess-help-command**. See Chapter 8 [Help], page 44 for more details.

ess-quit   [Command]

> *C-c C-q* Issue an exiting command to the inferior process, additionally also running **ess-cleanup** for disposing of any temporary buffers (such as help buffers and edit buffers) that may have been created. Use this command when you have finished your S session instead of simply quitting at the inferior process prompt, otherwise you will need to issue the command **ess-cleanup** explicitly to make sure that all the files that need to be saved have been saved, and that all the temporary buffers have been killed.

4.6 Is the Statistical Process running under ESS?

For the S languages (S, S-Plus, R) ESS sets an option in the current process that programs in the language can check to determine the environment in which they are currently running.

ESS sets **options(STERM="iESS")** for S language processes running in an inferior **iESS[S]** or **iESS[R]** buffer.

ESS sets `options(STERM="ddeESS")` for independent S-Plus for Windows processes running in the GUI and communicating with ESS via the DDE (Microsoft Dynamic Data Exchange) protocol through a `ddeESS[S]` buffer.

Other values of `options()$STERM` that we recommend are:

- `length`: Fixed length xterm or telnet window.
- `scrollable`: Unlimited length xterm or telnet window.
- `server`: S-Plus Stat Server.
- `BATCH`: BATCH.
- `Rgui`: R GUI.
- `Commands`: S-Plus GUI without DDE interface to ESS.

Additional values may be recommended in the future as new interaction protocols are created. Unlike the values `iESS` and `ddeESS`, ESS can't set these other values since the S language program is not under the control of ESS.

4.7 Using emacsclient

When starting R or S under Unix, ESS sets `options(editor="emacsclient")`. (Under Microsoft Windows, it will use gnuclient.exe rather than emacsclient, but the same principle applies.) Within your R session, for example, if you have a function called `iterator`, typing `fix(iterator)`, will show that function in a temporary Emacs buffer. You can then correct the function. When you kill the buffer, the definition of the function is updated. Using `edit()` rather than `fix()` means that the function is not updated. Finally, the S function `page(x)` will also show a text representation of the object `x` in a temporary Emacs buffer.

4.8 Other commands provided by inferior-ESS

The following commands are also available in the process buffer:

`comint-interrupt-subjob`   [Command]
>   *C-c C-c* Sends a Control-C signal to the ESS process. This has the effect of aborting the current command.

`ess-switch-to-inferior-or-script-buffer` *toggle-eob*   [Command]
>   *C-c C-z* When in process buffer, return to the most recent script buffer. When in a script buffer pop to the associated process buffer. This is a single key command, that is *C-c C-z C-z* from a script buffer returns to the original buffer.
>
>   If *toggle-eob* is given, the value of `ess-switch-to-end-of-proc-buffer` is toggled.

`ess-switch-to-end-of-proc-buffer`   [User Option]
>   If non-nil, `ess-switch-to-inferior-or-script-buffer` goes to end of process buffer.

Other commands available in Inferior S mode are discussed in Section "Shell Mode" in *The Gnu Emacs Reference Manual*.

5 Sending code to the ESS process

Other commands are also available for evaluating portions of code in the S process. These commands cause the selected code to be evaluated directly by the ESS process as if you had typed them in at the command line; the `source()` function is not used. You may choose whether both the commands and their output appear in the process buffer (as if you had typed in the commands yourself) or if the output alone is echoed. The behavior is controlled by the variable:

`ess-eval-visibly`   [User Option]
>   Non-`nil` means `ess-eval-*` commands display commands and output in the process buffer. Default is `t`.

Passing a prefix (`C-u`) *vis* to any of the following commands, however, reverses the meaning of `ess-eval-visibly` for that command only — for example `C-u C-c C-j` suppresses the current line of S (or other) code in the ESS process buffer. This method of evaluation is an alternative to S's `source()` function when you want the input as well as the output to be displayed. (You can sort of do this with `source()` when the option `echo=T` is set, except that prompts do not get displayed. ESS puts prompts in the right places.)

Primary commands for evaluating code are:

`ess-eval-region-or-line-and-step` *vis*   [Command]
>   Send the highlighted region or current line and step to next line of code.

`ess-eval-region-or-function-or-paragraph` *vis*   [Command]
>   `C-M-x` Sends the current selected region or function or paragraph.

`ess-eval-region-or-function-or-paragraph-and-step` *vis*   [Command]
>   `C-c C-c` Like `ess-eval-region-or-function-or-paragraph` but steps to next line of code.

Other, not so often used, evaluation commands are:

`ess-eval-line` *vis*   [Command]
>   `C-c C-j` Sends the current line to the ESS process.

`ess-eval-line-and-go` *vis*   [Command]
>   `C-c M-j` Like `ess-eval-line` but additionally switches point to the ESS process.

`ess-eval-function` *vis no-error*   [Command]
>   `C-c C-f` Sends the S function containing point to the ESS process.

`ess-eval-function-and-go` *vis*   [Command]
>   `C-c M-f` Like `ess-eval-function` but additionally switches point to the ESS process.

`ess-eval-region` *start end toggle message*   [Command]
>   `C-c C-r` Sends the current region to the ESS process.

`ess-eval-region-and-go` *start end vis*   [Command]
>   `C-c M-r` Like `ess-eval-region` but additionally switches point to the ESS process.

`ess-eval-buffer` *vis*   [Command]
> *C-c C-b* Sends the current buffer to the ESS process.

`ess-eval-buffer-and-go` *vis*   [Command]
> *C-c M-b* Like `ess-eval-buffer` but additionally switches point to the ESS process.

All the above `ess-eval-*` commands are useful for evaluating small amounts of code and observing the results in the process buffer for debugging purposes, or for generating transcripts from source files. When editing S functions, it is generally preferable to use *C-c C-l* to update the function's value. In particular, `ess-eval-buffer` is now largely obsolete.

A useful way to work is to divide the frame into two windows; one containing the source code and the other containing the process buffer. If you wish to make the process buffer scroll automatically when the output reaches the bottom of the window, you will need to set the variable `comint-scroll-to-bottom-on-output` to 'others or t.

6 Manipulating saved transcript files

Inferior S mode records the transcript (the list of all commands executed, and their output) in the process buffer, which can be saved as a *transcript file*, which should normally have the suffix `.St`. The most obvious use for a transcript file is as a static record of the actions you have performed in a particular S session. Sometimes, however, you may wish to re-execute commands recorded in the transcript file by submitting them to a running ESS process. This is what Transcript Mode is for.

If you load file a with the suffix `.St` into Emacs, it is placed in S Transcript Mode. Transcript Mode is similar to Inferior S mode (see Chapter 4 [Entering commands], page 26): paragraphs are defined as a command and its output, and you can move though commands either with the paragraph commands or with `C-c C-p` and `C-c C-n`.

6.1 Resubmitting commands from the transcript file

Three commands are provided to re-submit command lines from the transcript file to a running ESS process. They are:

ess-transcript-send-command   [Command]
> `M-RET` Sends the current command line to the ESS process, and execute it.

ess-transcript-copy-command   [Command]
> `C-c RET` Copy the current command to the ESS process, and switch to it (ready to edit the copied command).

ess-transcript-send-command-and-move   [Command]
> `RET` Sends the current command to the ESS process, and move to the next command line. This command is useful for submitting a series of commands.

Note that the first two commands are similar to those on the same keys in inferior S Mode. In all three cases, the commands should be executed when the cursor is on a command line in the transcript; the prompt is automatically removed before the command is submitted.

6.2 Cleaning transcript files

Yet another use for transcript files is to extract the command lines for inclusion in an S source file or function. Transcript mode provides one command which does just this:

ess-transcript-clean-region *beg end even-if-read-only*   [Command]
> `C-c C-w` Strip the transcript in the region (given by *beg* and *end*), leaving only commands. Deletes any lines not beginning with a prompt, and then removes the prompt from those lines that remain. Prefix argument *even-if-read-only* means to clean even if the buffer is read-only. Don't forget to remove any erroneous commands first!

The remaining command lines may then be copied to a source file or edit buffer for inclusion in a function definition, or may be evaluated directly (see Chapter 5 [Evaluating code], page 34) using the code evaluation commands from S mode, also available in S Transcript Mode.

7 Editing objects and functions

ESS provides facilities for editing S objects within your Emacs session. Most editing is performed on S functions, although in theory you may edit datasets as well. Edit buffers are always associated with files, although you may choose to make these files temporary if you wish. Alternatively, you may make use of a simple yet powerful mechanism for maintaining backups of text representations of S functions. Error-checking is performed when S code is loaded into the ESS process.

7.1 Creating or modifying S objects

To edit an S object, type

ess-dump-object-into-edit-buffer *object*   [Command]
    C-c C-e C-d Edit an S *object* in its own edit buffer.

from within the ESS process buffer (*S*). You will then be prompted for an object to edit: you may either type in the name of an existing object (for which completion is available using the *TAB* key), or you may enter the name of a new object. A buffer will be created containing the text representation of the requested object or, if you entered the name of a non-existent object at the prompt and the variable **ess-function-template** is non-**nil**, you will be presented with a template defined by that variable, which defaults to a skeleton function construct.

You may then edit the function as required. The edit buffer generated by **ess-dump-object-into-edit-buffer** is placed in the **ESS** major mode which provides a number of commands to facilitate editing S source code. Commands are provided to intelligently indent S code, evaluate portions of S code and to move around S code constructs.

Note: when you dump a file with *C-c C-e C-d*, ESS first checks to see whether there already exists an edit buffer containing that object and, if so, pops you directly to that buffer. If not, ESS next checks whether there is a file in the appropriate place with the appropriate name (see Section 7.6 [Source Files], page 40) and if so, reads in that file. You can use this facility to return to an object you were editing in a previous session (and which possibly was never loaded to the S session). Finally, if both these tests fail, the ESS process is consulted and a **dump()** command issued. If you want to force ESS to ask the ESS process for the object's definition (say, to reformat an unmodified buffer or to revert back to S's idea of the object's definition) pass a prefix argument to **ess-dump-object-into-edit-buffer** by typing *C-u C-c C-e C-d*.

7.2 Loading source files into the ESS process

The best way to get information — particularly function definitions — into S is to load them in as source file, using S's **source** function. You have already seen how to create source files using *C-c C-e C-d*; ESS provides a complementary command for loading source files (even files not created with ESS!) into the ESS process, namely **ess-load-file** (*C-c C-l*). see Section 4.5 [Hot keys], page 31.

After typing *C-c C-l* you will prompt for the name of the file to load into S; usually this is the current buffer's file which is the default value (selected by simply pressing *RET* at the prompt). You will be asked to save the buffer first if it has been modified (this happens automatically if the buffer was generated with *C-c C-e C-d*). The file will then be loaded, and if it loads successfully you will be returned to the ESS process.

7.3 Detecting errors in source files

If any errors occur when loading a file with `C-c C-l`, ESS will inform you of this fact. In this case, you can jump directly to the line in the source file which caused the error by typing *C-c '* (ess-parse-errors). You will be returned to the offending file (loading it into a buffer if necessary) with point at the line S reported as containing the error. You may then correct the error, and reload the file. Note that none of the commands in an S source file will take effect if any part of the file contains errors.

Sometimes the error is not caused by a syntax error (loading a non-existent file for example). In this case typing *C-c '* will simply display a buffer containing S's error message. You can force this behavior (and avoid jumping to the file when there *is* a syntax error) by passing a prefix argument to ess-parse-errors with *C-u C-c '*.

7.4 Indenting and formatting S code

ESS provides a sophisticated mechanism for indenting S source code (thanks to Ken'ichi Shibayama). Compound statements (delimited by '{' and '}') are indented relative to their enclosing block. In addition, the braces have been electrified to automatically indent to the correct position when inserted, and optionally insert a newline at the appropriate place as well. Lines which continue an incomplete expression are indented relative to the first line of the expression. Function definitions, `if` statements, calls to `expression()` and loop constructs are all recognized and indented appropriately. User variables are provided to control the amount of indentation in each case, and there are also a number of predefined indentation styles to choose from.

Comments are also handled specially by ESS, using an idea borrowed from the Emacs-Lisp indentation style. By default, comments beginning with '###' are aligned to the beginning of the line. Comments beginning with '##' are aligned to the current level of indentation for the block containing the comment. Finally, comments beginning with '#' are aligned to a column on the right (the 40th column by default, but this value is controlled by the variable `comment-column`,) or just after the expression on the line containing the comment if it extends beyond the indentation column. You turn off the default behavior by adding the line (setq ess-indent-with-fancy-comments nil) to your `.emacs` file.

ESS also supports Roxygen entries which is R documentation maintained in the source code as comments See Section 10.2.2 [Roxygen], page 52.

The indentation commands provided by ESS are:

ess-indent-or-complete   [Command]
> *TAB* Indents the current line as S code.
>
> Try to indent first, and if code is already properly indented, complete instead. In ess-mode, only tries completion if `ess-tab-complete-in-script` is non-nil. See also `ess-first-tab-never-complete`.

ess-tab-complete-in-script   [User Option]
> If non-`nil`, *TAB* in script buffers tries to complete if there is nothing to indent.

ess-first-tab-never-complete   [User Option]
> If non-nil, *TAB* never tries to complete in ess-mode. The default 'symbol does not
> try to complete if the next char is a valid symbol constituent. There are more options,
> see the help (*C-h v*).

ess-indent-exp   [Command]
> *TAB* Indents each line in the S (compound) expression which follows point. Very useful
> for beautifying your S code.

ess-electric-brace   [Command]
> { } The braces automatically indent to the correct position when typed.

The following Emacs command are also helpful:

RET
LFD   newline-and-indent Insert a newline, and indent the next line. (Note that
>   most keyboards nowadays do not have a LINEFEED key, but RET and *C-j* are
>   equivalent.)

M-;   indent-for-comment Indents an existing comment line appropriately, or inserts
>   an appropriate comment marker.

7.4.1 Changing indentation styles

The combined value of nine variables that control indentation are collectively termed a *style*.
ESS provides several styles covering the common styles of indentation: DEFAULT, OWN, GNU,
BSD, K&R, C++, RRR, CLB. The variable ess-style-alist lists the value of each indentation
variable per style.

ess-set-style   [Command]
> *C-c .* Sets the formatting style in this buffer to be one of the predefined styles,
> including GNU, BSD, K&R, CLB, and C++. The DEFAULT style uses the default values for
> the indenting variables; The OWN style allows you to use your own private values of
> the indentation variable, see below.
>
>   (setq ess-default-style 'C++)

ess-default-style   [User Option]
> The default value of ess-style. See the variable ess-style-alist for how these
> groups (DEFAULT, OWN, GNU, BSD, ...) map onto different settings for variables.

ess-style-alist   [User Option]
> Predefined formatting styles for ESS code. Values for all groups, except OWN, are
> fixed. To change the value of variables in the OWN group, customize the variable
> ess-own-style-list. The default style in use is controlled by ess-default-style.

The styles DEFAULT and OWN are initially identical. If you wish to edit some of the default
values, set ess-default-style to 'OWN and change ess-own-style-list. See Appendix A
[Customization], page 82, for convenient ways to set both these variables.

If you prefer not to use the custom facility, you can change individual indentation variables within a hook, for example:

```
(defun myindent-ess-hook ()
  (setq ess-indent-level 4))
(add-hook 'ess-mode-hook 'myindent-ess-hook)
```

In the rare case that you'd like to add an entire new indentation style of your own, copy the definition of `ess-own-style-list` to a new variable and ensure that the last line of the `:set` declaration calls `ess-add-style` with a unique name for your style (e.g. 'MINE). Finally, add (setq ess-default-style 'MINE) to use your new style.

7.5 Commands for motion, completion and more

A number of commands are provided to move across function definitions in the edit buffer:

`ess-goto-beginning-of-function-or-para`   [Command]
> *ESC C-a* aka *C-M-a* If inside a function go to the beginning of it, otherwise go to the beginning of paragraph.

`ess-goto-end-of-function-or-para`   [Command]
> *ESC C-e* aka *C-M-e* Move point to the end of the function containing point.

`ess-mark-function`   [Command]
> *ESC C-h* aka *C-M-h* Place point at the beginning of the S function containing point, and mark at the end.

Don't forget the usual Emacs commands for moving over balanced expressions and parentheses: See Section "Lists and Sexps" in *The GNU Emacs Reference Manual*.

Completion is provided in the edit buffer in a similar fashion to the process buffer: `TAB` first indents, and if there is nothing to indent, completes the object or file name; *M-?* lists file completions. See See Chapter 9 [Completion], page 46, for more.

Finally, *C-c C-z* (`ess-switch-to-inferior-or-script-buffer`) returns you to the `iESS` process buffer, if done from a script buffer, placing point at the end of the buffer. If this is done from the `iESS` process buffer, point is taken to the script buffer.

In addition some commands available in the process buffer are also available in the script buffer. You can still read help files with *C-c C-v*, edit another function with *C-c C-e C-d* and of course *C-c C-l* can be used to load a source file into S.

7.6 Maintaining S source files

Every edit buffer in ESS is associated with a *dump file* on disk. Dump files are created whenever you type *C-c C-e C-d* (`ess-dump-object-into-edit-buffer`), and may either be deleted after use, or kept as a backup file or as a means of keeping several versions of an S function.

`ess-delete-dump-files`   [User Option]
> If non-`nil`, dump files created with C-c C-e C-d are deleted immediately after they are created by the ess-process.

Since immediately after S dumps an object's definition to a disk file the source code on disk corresponds exactly to S's idea of the object's definition, the disk file isn't needed; deleting it now has the advantage that if you *don't* modify the file (say, because you just wanted to look at the definition of one of the standard S functions) the source dump file won't be left around when you kill the buffer. Note that this variable only applies to files generated with S's **dump** function; it doesn't apply to source files which already exist. The default value is **t**.

ess-keep-dump-files   [User Option]

> Variable controlling whether to delete dump files after a successful load. If 'nil': always delete. If 'ask', confirm to delete. If 'check', confirm to delete, except for files created with **ess-dump-object-into-edit-buffer**. Anything else, never delete. This variable only affects the behaviour of **ess-load-file**. Dump files are never deleted if an error occurs during the load.

After an object has been successfully (i.e. without error) loaded back into S with *C-c C-l*, the disk file again corresponds exactly (well, almost — see below) to S's record of the object's definition, and so some people prefer to delete the disk file rather than unnecessarily use up space. This option allows you to do just that.

If the value of **ess-keep-dump-files** is **t**, dump files are never deleted after they are loaded. Thus you can maintain a complete text record of the functions you have edited within ESS. Backup files are kept as usual, and so by using the Emacs numbered backup facility — see Section "Single or Numbered Backups" in *The Gnu Emacs Reference Manual*, you can keep a historic record of function definitions. Another possibility is to maintain the files with a version-control system such as RCS See Section "Version Control" in *The Gnu Emacs Reference Manual*. As long as a dump file exists in the appropriate place for a particular object, editing that object with *C-c C-e C-d* finds that file for editing (unless a prefix argument is given) — the ESS process is not consulted. Thus you can keep comments *outside* the function definition as a means of documentation that does not clutter the S object itself. Another useful feature is that you may format the code in any fashion you please without S re-indenting the code every time you edit it. These features are particularly useful for project-based work.

If the value of **ess-keep-dump-files** is nil, the dump file is always silently deleted after a successful load with *C-c C-l*. While this is useful for files that were created with *C-c C-e C-d* it also applies to any other file you load (say, a source file of function definitions), and so can be dangerous to use unless you are careful. Note that since **ess-keep-dump-files** is buffer-local, you can make sure particular files are not deleted by setting it to **t** in the Local Variables section of the file See Section "Local Variables in Files" in *The Gnu Emacs Reference Manual*.

A safer option is to set **ess-keep-dump-files** to **ask**; this means that ESS will always ask for confirmation before deleting the file. Since this can get annoying if you always want to delete dump files created with *C-c C-e C-d*, but not any other files, setting **ess-keep-dump-files** to **check** (the default value) will silently delete dump files created with *C-c C-e C-d* in the current Emacs session, but query for any other file. Note that in any case you will only be asked for confirmation once per file, and your answer is remembered for the rest of the Emacs session.

Note that in all cases, if an error (such as a syntax error) is detected while loading the file with *C-c C-l*, the dump file is *never* deleted. This is so that you can edit the file in a new Emacs session if you happen to quit Emacs before correcting the error.

Dump buffers are always autosaved, regardless of the value of **ess-keep-dump-files**.

7.7 Names and locations of dump files

Every dump file should be given a unique file name, usually the dumped object name with some additions.

`ess-dump-filename-template`   [User Option]

> Template for filenames of dumped objects. `%s` is replaced by the object name.

By default, dump file names are the user name, followed by '.' and the object and ending with '.S'. Thus if user `joe` dumps the object `myfun` the dump file will have name `joe.myfun.S`. The username part is included to avoid clashes when dumping into a publicly-writable directory, such as `/tmp`; you may wish to remove this part if you are dumping into a directory owned by you.

You may also specify the directory in which dump files are written:

`ess-source-directory`   [User Option]

> Directory name (ending in a slash) where S dump files are to be written.

By default, dump files are always written to `/tmp`, which is fine when `ess-keep-dump-files` is `nil`. If you are keeping dump files, then you will probably want to keep them somewhere in your home directory, say `~/S-source`. This could be achieved by including the following line in your `.emacs` file:

```
(setq ess-source-directory (expand-file-name "~/S-source/"))
```

If you would prefer to keep your dump files in separate directories depending on the value of some variable, ESS provides a facility for this also. By setting `ess-source-directory` to a lambda expression which evaluates to a directory name, you have a great deal of flexibility in selecting the directory for a particular source file to appear in. The lambda expression is evaluated with the process buffer as the current buffer and so you can use the variables local to that buffer to make your choice. For example, the following expression causes source files to be saved in the subdirectory `Src` of the directory the ESS process was run in.

```
(setq ess-source-directory
      (lambda ()
        (concat ess-directory "Src/")))
```

(`ess-directory` is a buffer-local variable in process buffers which records the directory the ESS process was run from.) This is useful if you keep your dump files and you often edit objects with the same name in different ESS processes. Alternatively, if you often change your S working directory during an S session, you may like to keep dump files in some subdirectory of the directory pointed to by the first element of the current search list. This way you can edit objects of the same name in different directories during the one S session:

```
(setq ess-source-directory
   (lambda ()
      (file-name-as-directory
       (expand-file-name (concat
                          (car ess-search-list)
                          "/.Src")))))
```

If the directory generated by the lambda function does not exist but can be created, you will be asked whether you wish to create the directory. If you choose not to, or the directory cannot be created, you will not be able to edit functions.

8 Reading help files

ESS provides an easy-to-use facility for reading S help files from within Emacs. From within the ESS process buffer or any ESS edit buffer, typing *C-c C-v* (ess-display-help-on-object) will prompt you for the name of an object for which you would like documentation. Completion is provided over all objects which have help files.

If the requested object has documentation, you will be popped into a buffer (named *help(*obj-name*)*) containing the help file. This buffer is placed in a special 'S Help' mode which disables the usual editing commands but which provides a number of keys for paging through the help file.

Help commands:

?   ess-describe-help-mode Pops up a help buffer with a list of the commands available in S help mode.

h   ess-display-help-on-object Pop up a help buffer for a different object.

Paging commands:

b
DEL scroll-down Move one page backwards through the help file.

SPC scroll-up Move one page forwards through the help file.

>
<   end-of-buffer Move to the beginning and end of the help file, respectively.

Section-based motion commands:

n
p   ess-skip-to-previous-section and ess-skip-to-next-section Move to the next and previous section header in the help file, respectively. A section header consists of a number of capitalized words, followed by a colon.

In addition, the *s* key followed by one of the following letters will jump to a particular section in the help file. Note that R uses capitalized instead of all-capital section headers, e.g., 'Description:' instead of 'DESCRIPTION:' and also only some versions of S(-plus) have sections 'BACKGROUND', 'BUGS', 'OPTIONAL ARGUMENTS', 'REQUIRED ARGUMENTS', and 'SIDE EFFECTS'.

Do use *s ?* to get the current list of active key bindings.

'a'   ARGUMENTS:

'b'   BACKGROUND:

'B'   BUGS:

'd'   DESCRIPTION:

'D'   DETAILS:

'e'   EXAMPLES:

'n'   NOTE:

'O'   OPTIONAL ARGUMENTS:

'R'	REQUIRED ARGUMENTS:
'r'	REFERENCES:
's'	SEE ALSO:
'S'	SIDE EFFECTS:
'u'	USAGE:
'v'	VALUE:
'<'	Jumps to beginning of file
'>'	Jumps to end of file
'?'	Pops up a help buffer with a list of the defined section motion keys.

Evaluation:

l   `ess-eval-line-and-step` Evaluates the current line in the ESS process, and moves to the next line. Useful for running examples in help files.

r   `ess-eval-region` Send the contents of the current region to the ESS process. Useful for running examples in help files.

Quit commands:

q   `ess-help-quit` Return to previously selected buffer, and bury the help buffer.

k   `kill-buffer` Return to previously selected buffer, and kills the help buffer.

x   `ess-kill-buffer-and-go` Return to the ESS process, killing this help buffer.

Miscellaneous:

i   `ess-display-index` Prompt for a package and display it's help index.

v   `ess-display-vignettes` Display all available vignettes.

w   `ess-display-help-in-browser` Display current help page with the web browser.

/   `isearch-forward` Same as *C-s*.

In addition, all of the ESS commands available in the edit buffers are also available in S help mode (see Section 7.1 [Edit buffer], page 37). Of course, the usual (non-editing) Emacs commands are available, and for convenience the digits and - act as prefix arguments.

If a help buffer already exists for an object for which help is requested, that buffer is popped to immediately; the ESS process is not consulted at all. If the contents of the help file have changed, you either need to kill the help buffer first, or pass a prefix argument (with *C-u*) to `ess-display-help-on-object`.

Help buffers are marked as temporary buffers in ESS, and are deleted when `ess-quit` or `ess-cleanup` are called.

Help buffers normally appear in another window within the current frame. If you wish help buffers to appear in their own frame (either one per help buffer, or one for all help buffers), you can customize the variable `ess-help-own-frame`.

9 Completion

9.1 Completion of object names

The `TAB` key is for completion. The value of the variable `ess-first-tab-never-complete` controls when completion is allowed controls when completion is allowed to occur. In `ess-mode` `TAB` first tries to indent, and if there is nothing to indent, complete the object name instead.

TAB   `comint-dynamic-complete` Complete the S object name or filename before point.

When the cursor is just after a partially-completed object name, pressing `TAB` provides completion in a similar fashion to `tcsh` except that completion is performed over all known S object names instead of file names. ESS maintains a list of all objects known to S at any given time, which basically consists of all objects (functions and datasets) in every attached directory listed by the `search()` command along with the component objects of attached data frames (if your version of S supports them).

For example, consider the three functions (available in Splus version 3.0) called `binomplot()`, `binom.test()` and `binomial()`. Typing *bin TAB* after the S prompt will insert the characters 'om', completing the longest prefix ('binom') which distinguishes these three commands. Pressing *TAB* once more provides a list of the three commands which have this prefix, allowing you to add more characters (say, '.') which specify the function you desire. After entering more characters pressing *TAB* yet again will complete the object name up to uniqueness, etc. If you just wish to see what completions exist without adding any extra characters, type *M-?*.

`ess-list-object-completions`   [Command]
     M-? List all possible completions of the object name at point.

ESS also provides completion over the components of named lists and environments (after '$'), S4 classes slots (after @), package and namespace objects (after :: and :::).

Completion is also provided over file names, which is particularly useful when using S functions such as `get()` or `scan()` which require fully expanded file names.

In the Inferior ESS buffer, if the cursor is not in a string and does not follow a (partial) object name, the `TAB` key has a third use: it expands history references. See Section 4.4 [History expansion], page 30.

Efficiency in completion is gained by maintaining a cache of objects currently known to S; when a new object becomes available or is deleted, only one component of the cache corresponding to the associated directory needs to be refreshed. If ESS ever becomes confused about what objects are available for completion (such as when if refuses to complete an object you **know** is there), the command *M-x ess-resynch* forces the *entire* cache to be refreshed, which should fix the problem.

9.2 Completion of function arguments

When inside a function call (i.e. following '('), *TAB* completion also provides function arguments. If function is a generic, completion will provide all the arguments of S3 methods known to R.

A related functionality is See Section 11.1 [ESS ElDoc], page 56, which displays function arguments in the echo area whenever the point is inside a function call.

9.3 Minibuffer completion

From version 12.03, ESS uses IDO mechanism (part of default emacs) for minibuffer completion if `ido.el` package is available and the value of `ess-use-ido` it `t` (the default). The completion command `ess-completing-read` falls back on classic `completion-read` interface if this feature is not available for whatever reason.

9.4 Integration with auto-complete package

ESS provides three sources for `Auto Completion` mode: `ac-source-R-args`, `ac-source-R-objects` and `ac-source-R`. The last one combines the previous two and makes them play nicely together. See auto-complete package documentation (http://cx4a.org/software/auto-complete/) for how to modify and install your own completion sources.

For the default auto-complete ESS configuration, install the latest version of `auto-complete` package. ESS automatically detect the package and activates auto-complete in ESS buffers.

To deactivate AC, place the following into your init file:

```
(setq ess-use-auto-complete nil)
```

Or, to activate auto-completion only in the `ess-mode` buffers:

```
(setq ess-use-auto-complete 'script-only)
```

ESS provides AC help both for arguments and objects (default keys *C-?* or *<f1>*). You can bind *M-h* to display quick help pop-ups:

```
(define-key ac-completing-map (kbd "M-h") 'ac-quick-help)
```

AC binds *M-n*, and *M-p* for the navigation in the completion menu, which might be inconvenient if you want it to use in the inferior R. Bind it to something else:

```
(define-key ac-completing-map "\M-n" nil) ;; was ac-next
(define-key ac-completing-map "\M-p" nil) ;; was ac-previous
(define-key ac-completing-map "\M-," 'ac-next)
(define-key ac-completing-map "\M-k" 'ac-previous)
```

9.5 Icicles

Another option for comprehensively handling completion in Emacs is via Icicles (http://www.emacswiki.org/emacs/Icicles). It allows users to have completions shown temporarily in the standard '*Completions*' buffer, and interactively select completion candidates using several methods. As of version 2013.04.04, Icicles provides support for completion in ESS. Please consult Icicles documentation, which is easily accessible via *customize-group Icicles*, for more details on installation and customization options.

Once installed, Icicles can be activated by evaluating (maybe place in `~/.emacs`):

```
(require 'icicles)
(icy-mode 1)
```

Icicles can be toggled at any moment by typing *M-x icy*.

When Icicles is on, *TAB* offers completion, provided the conditions determined by `ess-first-tab-never-complete` allow it. Typing *M-TAB* will attempt completion regardless. Typing *M-?* in ESS or iESS modes brings up the relevant completion candidates from which to choose. Minibuffer input filters the available candidates. Use *TAB* for prefix completion or *S-TAB* for substring or regexp completion. Use *S-SPC* to match an additional pattern (repeatable). You can cycle among the matching candidates, choosing with *RET* or *mouse-2*.

10 Developing with ESS

ESS provides several tools to help you with the development of your R packages:

10.1 ESS tracebug

ESS `tracebug` offers visual debugging, interactive error navigation, interactive backtrace, breakpoint manipulation, control over R error actions, watch window and interactive flagging/unflagging of functions for debugging.

From ESS13.05 `ess-tracebug` is on by default. You can toggle it on and off with *M-x* `ess-tracebug`. To disable startup activation of `ess-tracebug` set `ess-use-tracebug` to nil.

Tracebug functionality can be found on `ess-dev-map`, bound to *C-c C-t*. Additionally, when subprocess is in a debugging state **ess-debug-minor-mode** is active and the following additional shortcuts are available:

```
* Interactive Debugging ('ess-debug-minor-mode-map'):

   M-C    . Continue            . 'ess-debug-command-continue'
   M-C-C  . Continue multi      . 'ess-debug-command-continue-multi'
   M-N    . Next step           . 'ess-debug-command-next'
   M-C-N  . Next step multi     . 'ess-debug-command-next-multi'
   M-U    . Up frame            . 'ess-debug-command-up'
   M-Q    . Quit debugging      . 'ess-debug-command-quit'
```

These are all the tracebug commands defined in **ess-dev-map** (*C-c C-t ?* to show this table):

```
* Breakpoints ('ess-dev-map'):

   b   . Set BP (repeat to cycle BP type) . 'ess-bp-set'
   B   . Set conditional BP               . 'ess-bp-set-conditional'
   k   . Kill BP                          . 'ess-bp-kil'
   K   . Kill all BPs                     . 'ess-bp-kill-all'
   o   . Toggle BP state                  . 'ess-bp-toggle-state'
   l   . Set logger BP                    . 'ess-bp-set-logger'
   n   . Goto next BP                     . 'ess-bp-next'
   p   . Goto previous BP                 . 'ess-bp-previous'

   (C- prefixed equivalents are also defined)

* Debugging ('ess-dev-map'):
   '   . Show traceback                     . 'ess-show-traceback' (also on C-c ')■
   ~   . Show callstack                     . 'ess-show-call-stack' (also on C-c ~)■
   e   . Toggle error action (repeat to cycle). 'ess-debug-toggle-error-action'
   d   . Flag for debugging                 . 'ess-debug-flag-for-debugging'
   u   . Unflag for debugging               . 'ess-debug-unflag-for-debugging'■
```

```
w   . Watch window                          . 'ess-watch'

  (C- prefixed equivalents are also defined)

* Navigation to errors (general emacs functionality):

C-x ', M-g n   . 'next-error'
M-g p          . 'previous-error'

* Misc:

?   . Show this help           . 'ess-tracebug-show-help'
```

To configure how electric watch window splits the display see **ess-watch-width-threshold** and **ess-watch-height-threshold** variables.

A more detailed ess-tracebug documentation with screenshots is available at http://code.google.com/p/ess-tracebug/.

A short tutorial is at http://code.google.com/p/ess-tracebug/wiki/GettingStarted.

Note: Currently, ess-tracebug does not detect some of R's debug related messages in non-English locales. To set your R messages to English add the following line to your .Rprofile init file:

```
Sys.setlocale("LC_MESSAGES", "C")
```

10.2 Editing documentation

ESS provides two ways of writing documentation for R objects. Either using the standard R documentation system or using in-source documentation written as structured comment fields for use with the Roxygen package.

10.2.1 Editing R documentation (Rd) files

R objects are documented in files written in the *R documentation* ("Rd"), a simple markup language closely resembling (La)TEX, which can be processed into a variety of formats, including LaTEX, HTML, and plain text. Rd format is described in section "Rd format" of the "Writing R Extensions" manual in the R distribution. ESS has several features that facilitate editing Rd files.

Visiting an Rd file as characterized by its extension **Rd** will activate Rd Mode, which provides several facilities for making editing R documentation files more convenient, by helping with indentation, insertions, even doing some of the typing for you (with Abbrev Mode), and by showing Rd keywords, strings, etc. in different faces (with Font Lock Mode).

Note that R also accepts Rd files with extension **rd**; to activate ESS[Rd] support for this extension, you may need to add

```
(add-to-list 'auto-mode-alist '("\\.rd\\'" . Rd-mode))
```

to one of your Emacs startup files.

In Rd mode, the following special Emacs commands can be used in addition to the standard Emacs commands.

C-h m   Describe the features of Rd mode.

LFD

RET   Reindent the current line, insert a newline and indent the new line (`reindent-then-newline-and-indent`). An abbrev before point is expanded if `abbrev-mode` is non-`nil`.

TAB   Indent current line based on its contents and on previous lines. (`indent-according-to-mode`).

C-c C-e   Insert a "skeleton" with Rd markup for at least all mandatory entries in Rd files (`Rd-mode-insert-skeleton`). Note that many users might prefer to use the R function **prompt** on an existing R object to generate a non-empty Rd "shell" documenting the object (which already has all information filled in which can be obtained from the object).

C-c C-f   Insert "font" specifiers for some of the Rd markup commands markup available for emphasizing or quoting text, including markup for URLs and email addresses (`Rd-font`). *C-c C-f* is only a prefix; see e.g. *C-c C-f TAB* for the available bindings. Note that currently, not all of the Rd text markup as described in section "Marking text" of "Writing R Extensions" can be accessed via *C-c C-f*.

C-c C-j   Insert a suitably indented '\item{' on the next line (`Rd-mode-insert-item`).

C-c C-p   Preview a plain text version ("help file", see Chapter 8 [Help], page 44) generated from the Rd file (`Rd-preview-help`).

In addition, when editing Rd files one can interact with a running R process in a similar way as when editing R language files. E.g., *C-c C-v* provides access to on-line help, and *C-c C-n* sends the current line to the R process for evaluation. This interaction is particularly useful when editing the examples in the Rd file. See *C-h m* for all available commands.

Rd mode also provides access to abbreviations for most of the Rd markup commands. Type *M-x list-abbrevs* with Abbrev mode turned on to list all available abbrevs. Note that all Rd abbrevs start with a grave accent.

Rd mode can be customized via the following variables.

Rd-mode-hook   [User Option]
     Hook to be run when Rd mode is entered.

Rd-indent-level   [User Option]
     The indentation of Rd code with respect to containing blocks. Default is 2.

Rd-to-help-command   [User Option]
     The shell command used for converting Rd source to help text. Default is 'R CMD Rd2txt'.

To automatically turn on the abbrev and font-lock features of Rd mode, add the following lines to one of your Emacs startup files:

```
(add-hook 'Rd-mode-hook
          (lambda ()
           (abbrev-mode 1)
           (font-lock-mode 1)))
```

10.2.2 Editing Roxygen documentation

The Roxygen R package makes it possible to keep the intended contents for Rd files as structured comments in the R source files. Roxygen can then parse R files and generate appropriate Rd files from these comments, fill the usage fields as well as updating NAMESPACE files. See the Roxygen documentation found via http://roxygen.org for more information on its usage. An example of an Roxygen entry for a simple R function can look like this:

```
##' Description of the function
##'
##' Further details about this function
##' @title A title
##' @param me all parameters must be listed and documented
##' @return Description of the return value
##' @author The author
myfun <- function(me)
  cat("Hello", me, "\n")
```

The entry is immediately preceding the object to document and all lines start with the Roxygen prefix string, in this case ##'. ESS provides support to edit these documentation entries by providing line filling, navigation, template generation etc. Syntax highlighting is provided for Emacs but not for XEmacs.

Roxygen is customized by the variables in the customization group "Ess Roxy". Customizables include the Roxygen prefix, use of folding to toggle visibility of Roxygen entries and the Roxygen template.

All ESS Roxygen support is defined in ess-roxy.el which is loaded by default in ESS. The following special Emacs commands are provided.

ess-roxy-update-entry   [Command]
> *C-c C-o C-o* Generate a Roxygen template or update the parameter list in Roxygen entry at point (or above the function at the point). Documented parameters that are not in the function are placed last in the list, parameters that are not documented and not in the definition are dropped. Parameter descriptions are filled if ess-roxy-fill-param-p is non-nil.

ess-roxy-toggle-roxy-region *beg end*   [Command]
> *C-c C-o C-c* Toggle the presence of the leading Roxygen string on all lines in the marked region (between *beg* and *end*. Convenient for transferring text to Roxygen entries and to evaluate example fields.

ess-roxy-preview-Rd *name-file*   [Command]
> *C-c C-o C-r* Use the attached R process to parse the entry at point to obtain the Rd code. Convenient for previewing and checking syntax. When used with the prefix argument *name-file*, i.e. *C-u C-c C-e C-r*, place the content in a buffer associated with a Rd file with the same name as the documentation. Requires the Roxygen package to be installed.

ess-roxy-preview-HTML *visit-instead-of-open*   [Command]

> *C-c C-o C-t* Use the attached R process to parse the entry at to generate HTML for the entry and open it in a browser. When used with the prefix argument *visit-instead-of-open*, i.e. *C-u C-c C-e C-t*, visit the generated HTML file instead. Requires the Roxygen and tools packages to be installed.

ess-roxy-previous-entry   [Command]

> *C-c C-o p* Go to start of the Roxygen entry above point.

ess-roxy-next-entry   [Command]

> *C-c C-o n* Go to end of the Roxygen entry above point.

ess-roxy-hide-all   [Command]

> *C-c C-o C-h* Use the hideshow mode to fold away the visibility of all Roxygen entries. Hide-show support must be enabled for this binding to get defined.

ESS also advises the following standard editing functions in order to make Roxygen editing more intuitive:

TAB
: **ess-R-complete-object-name** Complete Roxygen tag at point. E.g. doing *TAB* when the point is at the end of @par completes to @param.

M-h
: **mark-paragraph** If the transient mark mode is active, place mark and point at start end end of the field at point and activate the mark.

TAB
: **ess-indent-command** If hide-show support is enabled, fold away the visibility of the Roxygen entry at point.

M-q
: **fill-paragraph** Fill the Roxygen field at point.

C-a
: **move-beginning-of-line** Move to the point directly to the right of the Roxygen start string.

RET
: **newline-and-indent** Insert a new line and the Roxygen prefix string.

10.3 ESS developer

Usual ESS evaluation commands, See Chapter 5 [Evaluating code], page 34, send portions of the current buffer for the evaluation in the current environment (usually **R_GlobalEnv**). Often, when developing packages with namespaces, it is necessary to evaluate code directly in the package's environment or its namespace. The **ess-developer** utility provides such a functionality with minimal disruption of the usual ESS work-flow.

To understand how ess-developer works you must be familiar with namespace system in R. In a nutshell, all objects defined in a package 'foo' are stored in an environment called 'namespace:foo'. Parent environment of 'namespace:foo' is an environment 'imports:foo' which contains copies of all objects from other packages which 'foo' imports. Parent environment of 'imports:foo' is the 'namespace:base'. Parent environment of 'namespace:base' is .GlobalEnv. Thus functions and methods stored in 'namespace:foo' see all the objects in .GlobalEnv unless shadowed by objects in 'imports:foo', 'namespace:base', or 'namespace:foo' itself. There is another environment associated with 'foo' - 'package:foo'. This environment stores *copies* of exported objects from 'namespace:foo' and is placed on the search() path, i.e. if 'foo' is loaded and if you start with .GlobalEnv and iteratively call parent.env() you will get eventually to 'package:foo'. Thus all methods and functions defined in .GlobalEnv can "see" objects in 'package:foo' environment. See also `http://cran.r-project.org/doc/manuals/R-ints.html#Namespaces`.

In order to use ess-developer you must add names of the packages that you are developing to `ess-developer-packages`. You can also do that interactively with *C-c C-t C-a*. To remove packages from `ess-developer-packages` use *C-c C-t C-r*. When developer mode is on, the process mode line indicator displays a small or capital letter "d".

If variable `ess-developer-activate-in-package` is `t` (the default) R-mode will check after visiting the file whether or not the file is part of the package. If visited file is part of a package listed in `ess-developer-packages`, developer mode is activated automatically.

Developer mode is usually activated on per-file basis and a small "d" appears in the modeline. You can also activate ess-developer for all buffers connected to current process. This is done by toggling **ess-developer** in subprocess buffer. In this case a big "D" will appear in the modeline.

ess-developer *val*   [Command]
> *C-c C-t C-t* Toggle developer mode on and off. If called from script buffer, toggle developer on file-by-file basis. When called from process buffer, toggle developer on per-process basis.

ess-developer-add-package *from-attached remove*   [Command]
> *C-c C-t C-a* Add a package to your development list (`ess-developer-packages`).

ess-developer-remove-package   [Command]
> *C-c C-t C-r* Remove a package from your development list.

When you add a package to `ess-developer-packages`, ESS will ask for loading command. By default there are two options `library` and `load_all` from `devtools` package. You can configure this behavior in `ess-developer-load-on-add-commands`. To explicitly load the package containing current file use *C-c C-t l*.

ess-developer-load-package   [Command]
> *C-c C-t l* Load package with `load_all` utility from `devtools` package.

When developer mode is on, ESS evaluation commands behave differently:

- *C-c C-l* (**ess-load-file**) asks for the package to source into and inserts all redefined objects into the package:foo or namespace:foo accordingly.
  - PLAIN OBJECTS and FUNCTIONS:
    If the object is found in an environment (package:foo or namespace:foo), and differs from the old one it is assigned into the corresponding environment. If the object is not found it is assigned into .GlobalEnv. The environment of functions is set to namespace:foo.
  - CLASSES:
    Same as plain objects, with the difference that even if the class definition is assigned into .GlobalEnv, it is still associated with the package foo. Thus if you issue getClassDeff("foo") you will get a class definition with the slot @package pointing to package "foo".

Note: Occasionally, after adding new classes you might get warnings from "setClass". This is especially true if new class inherits or is inherited by a class whose definition is not exported. You might get something like: `Warning: Class "boo" is defined (with package slot foo) but no metadata object found to revise subclass information---not exported?`

You can safely ignore this warnings.

- S4 METHODS:

  Similarly to function definitions modified methods are assigned in the local method table in the namespace:foo. New methods are assigned into .GlobalEnv, but with the environment pointing to namespace:foo. There is a subtle catch with method caching in R though. See the code in etc/ESSR/developer.R for more details.

- S3 METHODS:

  S3 methods are not automatically registered. You can register them manually after you have inserted method_name.my_class into your package environment using ess-developer, like follows:

  registerS3method("method_name", "my_class", my_package:::method_name.my_class)

  If you don't register your S3 method, R will call the registered (aka cached) S3 method instead of the new method that ess-developer inserted in the package environment.

- *C-c C-r* (`ess-eval-region`) and functions that depend on it (`ess-eval-paragraph`, `ess-eval-buffer` etc.), behave as `ess-load-file`, but restrict the evaluation to the corresponding region.

- *C-c C-f* (`ess-eval-function`) and friends check if the current function's name can be found in a namespace:foo or package:foo for a 'foo' from 'ess-developer-packages'. If found, and new function definition differs from the old one, the function is assigned into that namespace. If not found, it is assigned into .GlobalEnv.

11   Other ESS features and tools

ESS has a few extra features, which didn't fit anywhere else.

11.1   ElDoc

In `ElDoc` mode, the echo area displays function's arguments at point. From ESS version 12.03, ElDoc is active by default in `ess-mode` and `inferior-ess-mode` buffers. To activate it only in `ess-mode` buffers, place the following into your init file:

```
(setq ess-use-eldoc 'script-only)
```

`ess-use-eldoc`   [User Option]

>   If 't', activate eldoc in ess-mode and inferior-ess-mode buffers. If ''script-only' activate in ess-mode buffers only. Set `ess-use-eldoc` to `nil` to stop using `ElDoc` altogether.

`ess-eldoc-show-on-symbol`   [User Option]

>   This variable controls whether the help is shown only inside function calls. If set to 't', `ElDoc` shows help string whenever the point is on a symbol, otherwise (the default), shows only when the point is in a function call, i.e. after ''(''.

`ess-eldoc-abbreviation-style`   [User Option]

>   The variable determines how the doc string should be abbreviated to fit into minibuffer. Posible values are 'nil', 'mild', 'normal', 'strong' and 'aggressive'. Please see the documentation of the variable for more details. The default filter is 'normal'.

Ess-eldoc also honors the value of `eldoc-echo-area-use-multiline-p`, which if set to 'nil', will cause the truncation of doc string indifferent of the value of `ess-eldoc-abbreviation-style`. This way you can combine different filter levels with the truncation.

11.2   Handy commands

`ess-handy-commands`   [Command]

>   Request and execute a command from `ess-handy-commands` list.

`ess-handy-commands`   [User Option]

>   An alist of custom ESS commands available for call by `ess-handy-commands` and `ess-smart-comma` function.
>
>   Currently contains:
>
>   change-directory
>   `ess-change-directory`
>
>   help-index   `ess-display-index`
>
>   help-object
>   `ess-display-help-on-object`
>
>   vignettes   `ess-display-vignettes`
>
>   objects[ls]   `ess-execute-objects`

search	`ess-execute-search`
set-width	`ess-execute-screen-options`
install.packages	
	`ess-install.packages`
library	`ess-library`
setRepos	`ess-setRepositories`
sos	`ess-sos`

Handy commands: `ess-library`, `ess-install.packages`, etc - ask for item with completion and execute the correspond command. `ess-sos` is a interface to `findFn` function in package `sos`. If package `sos` is not found, ask user for interactive install.

11.3 Syntactic highlighting of buffers

ESS provides Font-Lock (see Section "Using Multiple Typefaces" in *The Gnu Emacs Reference Manual*) patterns for Inferior S Mode, S Mode, and S Transcript Mode buffers.

Syntax highlighting within ESS buffers is controlled by:

`ess-font-lock-mode` [User Option]

Non-'nil' means we use font lock support for ESS buffers. Default is 't', to use font lock support. If you change the value of this variable, restart Emacs for it to take effect.

The font-lock patterns are defined by the following variables, which you may modify if desired:

`inferior-R-font-lock-keywords` [User Option]

Font-lock patterns for inferior *R* processes. (There is a corresponding `inferior-S-font-lock-keywords` for *S* processes.) The default value highlights prompts, inputs, assignments, output messages, vector and matrix labels, and literals such as 'NA' and TRUE.

`ess-R-font-lock-keywords` [User Option]

Font-lock patterns for ESS R programming mode. (There is a corresponding `ess-S-font-lock-keywords` for S buffers.) The default value highlights function names, literals, assignments, source functions and reserved words.

11.4 Parenthesis matching

Emacs and XEmacs have facilities for highlighting the parenthesis matching the parenthesis at point. This feature is very useful when trying to examine which parentheses match each other. This highlighting also indicates when parentheses are not matching. Depending on what version of emacs you use, one of the following should work in your initialisation file:

```
(paren-set-mode 'paren) ;for XEmacs
(show-paren-mode t) ;for Emacs
```

11.5 Using graphics with ESS

One of the main features of the S package is its ability to generate high-resolution graphics plots, and ESS provides a number of features for dealing with such plots.

11.5.1 Using ESS with the `printer()` driver

This is the simplest (and least desirable) method of using graphics within ESS. S's `printer()` device driver produces crude character based plots which can be contained within the ESS process buffer itself. To start using character graphics, issue the S command

```
printer(width=79)
```

(the `width=79` argument prevents Emacs line-wrapping at column 80 on an 80-column terminal. Use a different value for a terminal with a different number of columns.) Plotting commands do not generate graphics immediately, but are stored until the `show()` command is issued, which displays the current figure.

11.5.2 Using ESS with windowing devices

Of course, the ideal way to use graphics with ESS is to use a windowing system. Under X Windows, or X11, this requires that the DISPLAY environment variable be appropriately set.

11.5.3 Java Graphics Device

S+6.1 and newer on Windows contains a java library that supports graphics. Send the commands:

```
library(winjava)
java.graph()
```

to start the graphics driver. This allows you to use ESS for both interaction and graphics within S-PLUS. (Thanks to Tim Hesterberg for this information.)

11.6 Imenu

Imenu is an Emacs tool for providing mode-specific buffer indexes. In some of the ESS editing modes (currently SAS and S), support for Imenu is provided. For example, in S mode buffers, the menubar should display an item called "Imenu-S". Within this menubar you will then be offered bookmarks to particular parts of your source file (such as the starting point of each function definition).

Imenu works by searching your buffer for lines that match what ESS thinks is the beginning of a suitable entry, e.g. the beginning of a function definition. To examine the regular expression that ESS uses, check the value of `imenu-generic-expression`. This value is set by various ESS variables such as `ess-imenu-S-generic-expression`.

11.7 Toolbar

The R and S editing modes have support for a toolbar. This toolbar provides icons to act as shortcuts for starting new S/R processes, or for evaluating regions of your source buffers. The toolbar should be present if your emacs can display images. See Appendix A [Customization], page 82, for ways to change the toolbar.

11.8 TAGS

The Emacs tags facility can be used to navigate around your files containing definitions of S functions. This facility is independent of ESS usage, but is written here since ESS users may wish to take advantage of TAGS facility. Read more about emacs tags in an emacs manual.

Etags, the program that generates the TAGS file, does not yet know the syntax to recognise function definitions in S files. Hence, you will need to provide a regexp that matches your function definitions. Here is an example call (broken over two lines; type as one line) that should be appropriate.

```
etags --language=none
--regex='/\([^ \t]+\)[ \t]*<-[ \t]*function/\1/' *.R
```

This will find entries in your source file of the form:

```
some.name <- function
```

with the function name starting in column 0. Windows users may need to change the single quotes to double quotes.

R version 2.9.0 introduced a front-end script for finding R tags, which calls the 'rtags()' function. By default, this script will recursively search the directories for relevant tags in R/C/Rd files. To use this script from the command line, try the following to get started:

```
R CMD rtags --help
```

For further details, see http://developer.r-project.org/rtags.html

11.9 Rdired

Ess-rdired provides a dired-like buffer for viewing, editing and plotting objects in your current R session. If you are used to using the dired (directory editor) facility in Emacs, this mode gives you similar functionality for R objects.

Start an R session with *M-x R* and then store a few variables, such as:

```
s <- sin(seq(from=0, to=8*pi, length=100))
x <- c(1, 4, 9)
y <- rnorm(20)
z <- TRUE
```

Then use *M-x ess-rdired* to create a buffer listing the objects in your current environment and display it in a new window:

```
       mode length
 s    numeric   100
 x    numeric     3
 y    numeric    20
 z    logical     1
```

Type *C-h m* or *?* to get a list of the keybindings for this mode. For example, with your point on the line of a variable, 'p' will plot the object, 'v' will view it, and 'd' will mark the object for deletion ('x' will actually perform the deletion).

11.10 Rutils

Ess-rutils builds up on ess-rdired, providing key bindings for performing basic R functions in the inferior-ESS process buffer, such as loading and managing packages, object manipulation (listing, viewing, and deleting), and alternatives to `help.start()` and `RSiteSearch()` that use the `browse-url` Emacs function. The library can be loaded using *M-x load-file*, but the easiest is to include:

> `(require 'ess-rutils)`

in your .emacs. Once R is started with *M-x R*, and if the value of the customizable variable `ess-rutils-keys` is true, several key bindings become available in iESS process buffers:

`ess-rutils-local-pkgs`   [Command]
> *C-c C-.* *l* List all packages in all available libraries.

`ess-rutils-repos-pkgs`   [Command]
> *C-c C-.* *r* List available packages from repositories listed by `getOptions(``repos'')` in the current R session.

`ess-rutils-update-pkgs` *lib repos*   [Command]
> *C-c C-.* *u* Update packages in a particular library *lib* and repository *repos*.

`ess-rutils-apropos`   [Command]
> *C-c C-.* *a* Search for a string using apropos.

`ess-rutils-rm-all`   [Command]
> *C-c C-.* *m* Remove all R objects.

`ess-rutils-objs`   [Command]
> *C-c C-.* *o* Manipulate R objects; wrapper for `ess-rdired`.

`ess-rutils-load-wkspc`   [Command]
> *C-c C-.* *w* Load a workspace file into R.

`ess-rutils-save-wkspc`   [Command]
> *C-c C-.* *s* Save a workspace file.

`ess-change-directory`   [Command]
> *C-c C-.* *d* Change the working directory for the current R session.

`ess-rutils-html-docs`   [Command]
> *C-c C-.* *H* Use `browse-url` to navigate R html documentation.

See the submenu 'Rutils' under the iESS menu for alternative access to these functions. The function `ess-rutils-rsitesearch` is provided without a particular key binding. This function is useful in any Emacs buffer, so can be bound to a user-defined key:

> `(eval-after-load "ess-rutils"`
>   `'(global-set-key [(control c) (f6)] 'ess-rutils-rsitesearch))`

Functions for listing objects and packages (`ess-rutils-local-pkgs`, `ess-rutils-repos-pkgs`, and `ess-rutils-objs`) show results in a separate buffer and window, in `ess-rutils-mode`, providing useful key bindings in this mode (type *?* in this buffer for a description).

11.11 Interaction with Org mode

Org-mode (http://orgmode.org) now supports reproducible research and literate programming in many languages (including R) – see chapter 14 of the Org manual (http://orgmode.org/org.html#Working-With-Source-Code). For ESS users, this offers a document-based work environment within which to embed ESS usage. R code lives in code blocks of an Org document, from which it can be edited in ess-mode, evaluated, extracted ("tangled") to pure code files. The code can also be exported ("woven") with the surrounding text to several formats including HTML and LaTeX. Results of evaluation including figures can be captured in the Org document, and data can be passed from the Org document (e.g. from a table) to the ESS R process. (This section contributed by Dan Davison and Eric Schulte.)

11.12 Support for Sweave in ESS and AUCTeX

ESS provides support for writing and processing Sweave (http://www.statistik.lmu.de/~leisch/Sweave), building up on Emacs' ess-noweb-mode for literate programming. When working on an Sweave document, the following key bindings are available:

ess-swv-weave *choose* [Command]

    M-n s Run Sweave on the current .Rnw file. If *choose* is non-'nil', offer a menu of available weavers.

ess-swv-latex [Command]

    M-n l Run LaTeX after Sweave'ing.

ess-swv-PS [Command]

    M-n p Generate and display a postscript file after LaTeX'ing.

ess-swv-PDF *pdflatex-cmd* [Command]

    M-n P Generate and display a PDF file after LaTeX'ing. Optional argument *pdflatex-cmd* is the command to use, which by default, is the command used to generate the PDF file is the first element of **ess-swv-pdflatex-commands**.

ess-swv-pdflatex-commands [User Option]

    Commands used by **ess-swv-PDF** to run a version of pdflatex; the first entry is the default command.

Sweave'ing with **ess-swv-weave** starts an inferior-ESS process, if one is not available. Other commands are available from the 'Sweaving, Tangling, ...' submenu of the Noweb menu.

AUCTeX (http://www.gnu.org/software/auctex) users may prefer to set the variable **ess-swv-plug-into-AUCTeX-p** (available from the "ESS Sweave" customization group) to t. Alternatively, the same can be achieved by activating the entry "AUCTeX Interface" from the 'Sweaving, Tangling, ...' submenu, which toggles this variable on or off. When the interface is activated, new entries for Sweave'ing and LaTeX'ing thereafter are available from AUCTeX's "Command" menu. Sweave'ing can, thus, be done by *C-c C-c Sweave RET* without an inferior-ESS process. Similarly, LaTeX'ing can be done by *C-c C-c LaTeXSweave RET*. In both cases, the process can be monitored with *C-c C-l* (TeX-recenter-output-buffer). Open the viewer with *C-c C-v* (TeX-view), as usual in AUCTeX.

12 Overview of ESS features for the S family

12.1 ESS[S]–Editing files

ESS[S] is the mode for editing S language files. This mode handles:

- proper indenting, generated by both [Tab] and [Return].
- color and font choices based on syntax.
- ability to send the contents of an entire buffer, a highlighted region, an S function, or a single line to an inferior S process, if one is currently running.
- ability to switch between processes which would be the target of the buffer (for the above).
- The ability to request help from an S process for variables and functions, and to have the results sent into a separate buffer.
- completion of object names and file names.

ESS[S] mode should be automatically turned on when loading a file with the suffices found in ess-site (*.R, *.S, *.s, etc). Alternatively, type *M-x R-mode* to put the current buffer into R mode. However, one will have to start up an inferior process to take advantage of the interactive features.

12.2 iESS[S]–Inferior ESS processes

iESS (inferior ESS) is the mode for interfacing with active statistical processes (programs). This mode handles:

- proper indenting, generated by both [Tab] and [Return].
- color and font highlighting based on syntax.
- ability to resubmit the contents of a multi-line command to the executing process with a single keystroke [RET].
- The ability to request help from the current process for variables and functions, and to have the results sent into a separate buffer.
- completion of object names and file names.
- interactive history mechanism.
- transcript recording and editing.

To start up iESS mode, use:

```
M-x S+3
M-x S4
M-x S+5
M-x S+6
M-x R
```

(for S-PLUS 3.x, S4, S+5, S+6 or S+7, and R, respectively. This assumes that you have access to each). Usually the site will have defined one of these programs (by default S+6) to the simpler name:

M-x S

In the (rare) case that you wish to pass command line arguments to the starting S+6 process, set the variable `inferior-Splus-args`.

Note that R has some extremely useful command line arguments. For example, `--vanilla` will ensure R starts up without loading in any init files. To enter a command line argument, call R using a "prefix argument", by

C-u M-x R

and when ESS prompts for "Starting Args ? ", enter (for example):

`--vanilla`

Then that R process will be started up using `R --vanilla`. If you wish to always call R with certain arguments, set the variable `inferior-R-args` accordingly.

If you have other versions of R or S-Plus available on the system, ESS is also able to start those versions. How this exactly works depend on which OS you are using, as described in the following paragraphs. The general principle, regardless of OS, is that ESS searches the paths listed in the variable `exec-path` for R binaries. If ESS cannot find your R binaries, on Unix you can change the unix environment variable `PATH`, as this variable is used to set `exec-path`.

R on Unix systems: If you have "R-1.8.1" on your `exec-path`, it can be started using *M-x R-1.8.1*. By default, ESS will find versions of R beginning "R-1" or "R-2". If your versions of R are called other names, consider renaming them with a symbolic link or change the variable `ess-r-versions`. To see which functions have been created for starting different versions of R, type *M-x R-* and then hit [Tab]. These other versions of R can also be started from the "ESS->Start Process->Other" menu.

R on Windows systems: If you have "rw1081" on your `exec-path`, it can be started using *M-x rw1081*. By default, ESS will find versions of R located in directories parallel to the version of R in your `PATH`. If your versions of R are called other names, you will need to change the variable `ess-rterm-versions`. To see which functions have been created for starting different versions of R, type *M-x rw* and then hit [Tab]. These other versions of R can also be started from the "ESS->Start Process->Other" menu.

Once ESS has found these extra versions of R, it will then create a new function, called *M-x R-newest*, which will call the newest version of R that it found. (ESS examines the date in the first line of information from `R --version` to determine which is newest.)

S on Unix systems: If you have "Splus7" on your `exec-path`, it can be started using *M-x Splus7*. By default, ESS will find all executables beginning "Splus" on your path. If your versions of S are called other names, consider renaming them with a symbolic link or change the variable `ess-s-versions`. To see which functions have been created for starting different versions of Splus, type *M-x Splus* and then hit [Tab]. These other versions of Splus can also be started from the "ESS->Start Process->Other" menu.

A second mechanism is also available for running other versions of Splus. The variable `ess-s-versions-list` is a list of lists; each sublist should be of the form: (DEFUN-NAME PATH ARGS). DEFUN-NAME is the name of the new emacs function you wish to create to start the new S process; PATH is the full path to the version of S you want to run; ARGS is an optional string of command-line arguments to pass to the S process. Here is an example setting:

```
(setq ess-s-versions-list
```

```
'( ("Splus64" "/usr/local/bin/Splus64")
   ("Splus64-j" "/usr/local/bin/Splus64" "-j")))
```

which will then allow you to do *M-x Splus64-j* to start Splus64 with the corresponding command line arguments.

If you change the value of either `ess-s-versions` or `ess-s-versions-list`, you should put them in your .emacs before ess-site is loaded, since the new emacs functions are created when ESS is loaded.

Sqpe (S-Plus running inside an emacs buffer) on Windows systems: If you have an older version of S-Plus (S-Plus 6.1 for example) on your system, ir can be started inside an emacs buffer with *M-x splus61*. By default, ESS will find versions of S-Plus located in the installation directories that Insightful uses by default. If your versions of S-Plus are anywhere else, you will need to change the variable `ess-SHOME-versions`. To see which functions have been created for starting different versions of S-Plus, type *M-x spl* and then hit [Tab]. These other versions of S-Plus can also be started from the "ESS->Start Process->Other" menu.

12.3 ESS-help–assistance with viewing help

ESS has built-in facilities for viewing help files from S. See Chapter 8 [Help], page 44.

12.4 Philosophies for using ESS[S]

The first is preferred, and configured for. The second one can be retrieved again, by changing emacs variables.

1: (preferred by the current group of developers): The source code is real. The objects are realizations of the source code. Source for EVERY user modified object is placed in a particular directory or directories, for later editing and retrieval.

2: (older version): S objects are real. Source code is a temporary realization of the objects. Dumped buffers should not be saved. _We_strongly_discourage_this_approach_. However, if you insist, add the following lines to your .emacs file:

```
(setq ess-keep-dump-files 'nil)
(setq ess-delete-dump-files t)
(setq ess-mode-silently-save nil)
```

The second saves a small amount of disk space. The first allows for better portability as well as external version control for code.

12.5 Scenarios for use (possibilities–based on actual usage)

We present some basic suggestions for using ESS to interact with S. These are just a subset of approaches, many better approaches are possible. Contributions of examples of how you work with ESS are appreciated (especially since it helps us determine priorities on future enhancements)! (comments as to what should be happening are prefixed by "##").

1: ##   Data Analysis Example (source code is real)
Load the file you want to work with
C-x C-f myfile.s

Edit as appropriate, and then start up S-PLUS 3.x
M-x S+3

A new buffer *S+3:1* will appear. Splus will have been started
in this buffer. The buffer is in iESS [S+3:1] mode.

Split the screen and go back to the file editing buffer.
C-x 2 C-x b myfile.s

Send regions, lines, or the entire file contents to S-PLUS. For regions,
highlight a region with keystrokes or mouse and then send with:
C-c C-r

Re-edit myfile.s as necessary to correct any difficulties. Add
new commands here. Send them to S by region with C-c C-r, or
one line at a time with C-c C-n.

Save the revised myfile.s with C-x C-s.

Save the entire *S+3:1* interaction buffer with C-c C-s. You
will be prompted for a file name. The recommended name is
myfile.St. With the *.St suffix, the file will come up in ESS
Transcript mode the next time it is accessed from Emacs.

2: ## Program revision example (source code is real)

Start up S-PLUS 3.x in a process buffer (this will be *S+3:1*)
M-x S+3

Load the file you want to work with
C-x C-f myfile.s

edit program, functions, and code in myfile.s, and send revised
functions to S when ready with
C-c C-f
or highlighted regions with
C-c C-r
or individual lines with
C-c C-n
or load the entire buffer with
C-c C-l

save the revised myfile.s when you have finished
C-c C-s

3: ## Program revision example (S object is real)

   ## Start up S-PLUS 3.x in a process buffer (this will be *S+3:1*)
   M-x S+3

   ## Dump an existing S object my.function into a buffer to work with
   C-c C-d my.function
   ## a new buffer named yourloginname.my.function.S will be created with
   ## an editable copy of the object. The buffer is associated with the
   ## pathname /tmp/yourloginname.my.function.S and will amlost certainly not
   ## exist after you log off.

   ## enter program, functions, and code into work buffer, and send
   ## entire contents to S-PLUS when ready
   C-c C-b

   ## Go to *S+3:1* buffer, which is the process buffer, and examine
   ## the results.
   C-c C-y
   ## The sequence C-c C-y is a shortcut for: C-x b *S+3:1*

   ## Return to the work buffer (may/may not be prefixed)
   C-x C-b yourloginname.my.function.S
   ## Fix the function that didn't work, and resubmit by placing the
   ## cursor somewhere in the function and
   C-c C-f
   ## Or you could've selected a region (using the mouse, or keyboard
   ## via setting point/mark) and
   C-c C-r
   ## Or you could step through, line by line, using
   C-c C-n
   ## Or just send a single line (without moving to the next) using
   C-c C-j
   ## To fix that error in syntax for the "rchisq" command, get help
   ## by
   C-c C-v rchisq

4:   Data Analysis (S object is real)
     ## Start up S-PLUS 3.x, in a process buffer (this will be *S+3:1*)
     M-x S+3

     ## Work in the process buffer. When you find an object that needs
     ## to be changed (this could be a data frame, or a variable, or a
     ## function), dump it to a buffer:

C-c C-d my.cool.function

Edit the function as appropriate, and dump back in to the
process buffer
C-c C-b

Return to the S-PLUS process buffer
C-c C-y
Continue working.

When you need help, use
C-c C-v rchisq
instead of entering: help("rchisq")

12.6 Customization Examples and Solutions to Problems

1. Suppose that you are primarily an SPLUS 3.4 user, occasionally using S version 4, and sick and tired of the buffer-name *S+3* we've stuck you with. Simply edit the "ess-dialect" alist entry in the ess-sp3-d.el and ess-s4-d.el files to be "S" instead of "S4" and "S+3". This will ensure that all the inferior process buffer names are "*S*".

2. Suppose that you WANT to have the first buffer name indexed by ":1", in the same manner as your S-PLUS processes 2,3,4, and 5 (for you heavy simulation people). Then add after your (require 'ess-site) or (load "ess-site") command in your .emacs file, the line:

```
(setq ess-plain-first-buffername nil)
```

3. Fontlocking sometimes fails to behave nicely upon errors. When Splus dumps, a mis-matched " (double-quote) can result in the wrong font-lock face being used for the remainder of the buffer.

Solution: add a " at the end of the "Dumped..." statement, to revert the font-lock face back to normal.

4. When you change directory within a *R* or *S* session using the setwd() command, emacs does not recognise that you have changed the current directory.

Solution: Use *M-x ess-change-directory*. This will prompt you for the directory to change to. It will then change directory within the *S* buffer, and also update the emacs variable default-directory. Alternatively, if you have already executed setwd(), press *M-RET* within the *S* buffer so that Emacs can update default-directory.

13 ESS for SAS

ESS[SAS] was designed for use with SAS. It is descended from emacs macros developed by John Sall for editing SAS programs and **SAS-mode** by Tom Cook. Those editing features and new advanced features are part of ESS[SAS]. The user interface of ESS[SAS] has similarities with ESS[S] and the SAS Display Manager.

13.1 ESS[SAS]–Design philosophy

ESS[SAS] was designed to aid the user in writing and maintaining SAS programs, such as **foo.sas**. Both interactive and batch submission of SAS programs is supported.

ESS[SAS] was written with two primary goals.

1. The emacs text editor provides a powerful and flexible development environment for programming languages. These features are a boon to all programmers and, with the help of ESS[SAS], to SAS users as well.

2. Although a departure from SAS Display Manager, ESS[SAS] provides similar key definitions to give novice ESS[SAS] users a head start. Also, inconvenient SAS Display Manager features, like remote submission and syntax highlighting, are provided transparently; appealing to advanced ESS[SAS] users.

13.2 ESS[SAS]–Editing files

ESS[SAS] is the mode for editing SAS language files. This mode handles:

- proper indenting, generated by both **TAB** and **RET**.
- color and font choices based on syntax.
- ability to save and submit the file you are working on as a batch SAS process with a single keypress and to continue editing while it is runs in the background.
- capability of killing the batch SAS process through the ***shell*** buffer or allow the SAS process to keep on running after you exit emacs.
- single keypress navigation of **.sas**, **.log** and **.lst** files (**.log** and **.lst** files are refreshed with each keypress).
- ability to send the contents of an entire buffer, a highlighted region, or a single line to an interactive SAS process.
- ability to switch between processes which would be the target of the buffer (for the above).

ESS[SAS] is automatically turned on when editing a file with a **.sas** suffix (or other extension, if specified via **auto-mode-alist**). The function keys can be enabled to use the same function keys that the SAS Display Manager does. The interactive capabilities of ESS require you to start an inferior SAS process with *M-x SAS* (See Section 13.6 [iESS(SAS)– Interactive SAS processes], page 74.)

At this writing, the indenting and syntax highlighting are generally correct. Known issues: for multiple line * or %* comments, only the first line is highlighted; for **.log** files, only the first line of a **NOTE:**, **WARNING:** or **ERROR:** message is highlighted; unmatched single/double quotes in **CARDS** data lines are **NOT** ignored; in an iterative **DO** statement, **TO** and **BY** are not highlighted.

13.3 ESS[SAS]–TAB key

Two options. The TAB key is bound by default to sas-indent-line. This function is used to syntactically indent SAS code so PROC and RUN are in the left margin, other statements are indented sas-indent-width spaces from the margin, continuation lines are indented sas-indent-width spaces in from the beginning column of that statement. This is the type of functionality that emacs provides in most programming language modes. This functionality is activated by placing the following line in your initialization file prior to a require/load:

```
(setq ess-sas-edit-keys-toggle nil)
```

ESS provides an alternate behavior for TAB that makes it behave as it does in SAS Display Manager, i.e. move the cursor to the next stop. The alternate behavior also provides a "TAB" backwards, *C-TAB*, that moves the cursor to the stop to the left and deletes any characters between them. This functionality is obtained by placing the following line in your initialization file prior to a require/load:

```
(setq ess-sas-edit-keys-toggle t)
```

Under the alternate behavior, TAB is bound to *M-x tab-to-tab-stop* and the stops are defined by ess-sas-tab-stop-list.

13.4 ESS[SAS]–Batch SAS processes

Submission of a SAS batch job is dependent on your environment. ess-sas-submit-method is determined by your operating system and your shell. It defaults to 'sh unless you are running Windows or Mac Classic. Under Windows, it will default to 'sh if you are using a UNIX-imitating shell; otherwise 'ms-dos for an MS-DOS shell. On Mac OS X, it will default to 'sh, but under Mac Classic, it defaults to 'apple-script. You will also set this to 'sh if the SAS batch job needs to run on a remote machine rather than your local machine. This works transparently if you are editing the remote file via ange-ftp/EFS or tramp. Note that ess-sas-shell-buffer-remote-init is a Local Variable that defaults to "ssh" which will be used to open the buffer on the remote host and it is assumed that no password is necessary, i.e. you are using ssh-agent/ssh-add or the equivalent (see the discussion about Local Variables below if you need to change the default).

However, if you are editing the file locally and transferring it back and forth with Kermit, you need some additional steps. First, start Kermit locally before remotely logging in. Open a local copy of the file with the **ess-kermit-prefix** character prepended (the default is "#"). Execute the command *M-x ess-kermit-get* which automatically brings the contents of the remote file into your local copy. If you transfer files with Kermit manually in a *shell* buffer, then note that the Kermit escape sequence is *C-q C- c* rather than *C- c* which it would be in an ordinary terminal application, i.e. not in an emacs buffer. Lastly, note that the remote Kermit command is specified by **ess-kermit-command**.

The command used by the SUBMIT function key (F3 or F8) to submit a batch SAS job, whether local or remote, is ess-sas-submit-command which defaults to sas-program. sas-program is "invoke SAS using program file" for Mac Classic and "sas" otherwise. However, you may have to alter ess-sas-submit-command for a particular program, so it is defined as buffer-local. Conveniently, it can be set at the end of the program:

```
endsas;
```

```
Local variables:
ess-sas-submit-command: "sas8"
End:
```

The command line is also made of `ess-sas-submit-pre-command`, `ess-sas-submit-post-command` and `ess-sas-submit-command-options` (the last of which is also buffer-local). Here are some examples for your `~/.emacs` or `~/.xemacs/init.el` file (you may also use *M-x customize-variable*):

```
;'sh default
(setq ess-sas-submit-pre-command "nohup")
;'sh default
(setq ess-sas-submit-post-command "-rsasuser &")
;'sh example
(setq-default ess-sas-submit-command "/usr/local/sas/sas")
;'ms-dos default
(setq ess-sas-submit-pre-command "start")
;'ms-dos default
(setq ess-sas-submit-post-command "-rsasuser -icon")
;Windows example
(setq-default ess-sas-submit-command "c:/progra~1/sas/sas.exe")
;Windows example
(setq-default ess-sas-submit-command "c:\\progra~1\\sas\\sas.exe")
```

There is a built-in delay before a batch SAS job is submitted when using a UNIX-imitating shell under Windows. This is necessary in many cases since the shell might not be ready to receive a command. This delay is currently set high enough so as not to be a problem. But, there may be cases when it needs to be set higher, or could be set much lower to speed things up. You can over-ride the default in your `~/.emacs` or `~/.xemacs/init.el` file by:

```
(setq ess-sleep-for 0.2)
```

For example, `(setq ess-sas-global-unix-keys t)` keys shown, `(setq ess-sas-global-pc-keys t)` in parentheses; ESS[SAS] function keys are presented in the next section. Open the file you want to work with *C-x C-f foo.sas*. *foo*.sas will be in ESS[SAS] mode. Edit as appropriate, then save and submit the batch SAS job.

F3 (F8)

The job runs in the `*shell*` buffer while you continue to edit *foo*.sas. If `ess-sas-submit-method` is `'sh`, then the message buffer will display the shell notification when the job is complete. The `'sh` setting also allows you to terminate the SAS batch job before it is finished.

F8 (F3)

Terminating a SAS batch in the `*shell*` buffer.

kill *PID*

You may want to visit the `.log` (whether the job is still running or it is finished) and check for error messages. The `.log` will be refreshed and you will be placed in it's buffer. You will be taken to the first error message, if any.

F5 (F6)

Goto the next error message, if any.

```
F5 (F6)
```

Now, 'refresh' the .lst and go to it's buffer.

```
F6 (F7)
```

If you wish to make changes, go to the .sas file with.

```
F4 (F5)
```

Make your editing changes and submit again.

```
F3 (F8)
```

13.5 ESS[SAS]–Function keys for batch processing

The setup of function keys for SAS batch processing is unavoidably complex, but the usage of function keys is simple. There are five distinct options:

Option 1 (default). Function keys in ESS[SAS] are not bound to elisp commands. This is in accordance with the GNU Elisp Coding Standards (GECS) which do not allow function keys to be bound so that they are available to the user.

Options 2-5. Since GECS does not allow function keys to be bound by modes, these keys are often unused. So, ESS[SAS] provides users with the option of binding elisp commands to these keys. Users who are familiar with SAS will, most likely, want to duplicate the function key capabilities of the SAS Display Manager. There are four options (noted in parentheses).

a. SAS Display Manager has different function key definitions for UNIX (2, 4) and Windows (3, 5); ESS[SAS] can use either.

b. The ESS[SAS] function key definitions can be active in all buffers (global: 4, 5) or limited (local: 2, 3) only to buffers with files that are associated with ESS[SAS] as specified in your auto-mode-alist.

The distinction between local and global is subtle. If you want the ESS[SAS] definitions to work when you are in the *shell* buffer or when editing files other than the file extensions that ESS[SAS] recognizes, you will most likely want to use the global definitions. If you want your function keys to understand SAS batch commands when you are editing SAS files, and to behave normally when editing other files, then you will choose the local definitions. The option can be chosen by the person installing ESS for a site or by an individual.

a. For a site installation or an individual, place **ONLY ONE** of the following lines in your initialization file prior to a require/load. ESS[SAS] function keys are available in ESS[SAS] if you choose either 2 or 3 and in all modes if you choose 4 or 5:

```
;;2; (setq ess-sas-local-unix-keys t)
;;3; (setq ess-sas-local-pc-keys t)
;;4; (setq ess-sas-global-unix-keys t)
;;5; (setq ess-sas-global-pc-keys t)
```

The names -unix- and -pc- have nothing to do with the operating system that you are running. Rather, they mimic the definitions that the SAS Display Manager uses by default on those platforms.

b. If your site installation has configured the keys contrary to your liking, then you must call the appropriate function.

```
(load "ess-site") ;; local-unix-keys
(ess-sas-global-pc-keys)
```

Finally, we get to what the function keys actually do. You may recognize some of the nicknames as SAS Display Manager commands (they are in all capitals).

UNIX	PC	Nickname
F2	F2	'refresh'
		revert the current buffer with the file of the same name if the file is newer than the buffer
F3	F8	SUBMIT
		save the current .sas file (which is either the .sas file in the current buffer or the .sas file associated with the .lst or .log file in the current buffer) and submit the file as a batch SAS job
F4	F5	PROGRAM
		switch buffer to .sas file
F5	F6	LOG
		switch buffer to .log file, 'refresh' and goto next error message, if any
F6	F7	OUTPUT
		switch buffer to .lst file and 'refresh'
F7	F4	'filetype-1'
		switch buffer to 'filetype-1' (defaults to .txt) file and 'refresh'
F8	F3	'shell'
		switch buffer to *shell*
F9	F9	VIEWTABLE
		open an interactive PROC FSEDIT session on the SAS dataset near point
F10	F10	toggle-log
		toggle ESS[SAS] for .log files; useful for certain debugging situations
F11	F11	'filetype-2'
		switch buffer to 'filetype-2' (defaults to .dat) file and 'refresh'
F12	F12	viewgraph
		open a GSASFILE near point for viewing either in emacs or with an external viewer
C-F1	C-F1	rtf-portrait
		create an MS RTF portrait file from the current buffer with a file extension of .rtf
C-F2	C-F2	rtf-landscape
		create an MS RTF landscape file from the current buffer with a file extension of .rtf
C-F3	C-F8	submit-region
		write region to ess-temp.sas and submit
C-F5	C-F6	append-to-log
		append ess-temp.log to the current .log file
C-F6	C-F7	append-to-output
		append ess-temp.lst to the current .lst file
C-F9	C-F9	INSIGHT
		open an interactive PROC INSIGHT session on the SAS dataset near point

C-F10   C-F10   toggle-listing

                      toggle ESS[SAS] for `.lst` files; useful for toggling read-only

`SUBMIT`, `PROGRAM`, `LOG` and `OUTPUT` need no further explanation since they mimic the SAS Display Manager commands and related function key definitions. However, six other keys have been provided for convenience and are described below.

'`shell`' switches you to the `*shell*` buffer where you can interact with your operating system. This is especially helpful if you would like to kill a SAS batch job. You can specify a different buffer name to associate with a SAS batch job (besides `*shell*`) with the buffer-local variable `ess-sas-shell-buffer`. This allows you to have multiple buffers running SAS batch jobs on multiple local/remote computers that may rely on different methods specified by the buffer-local variable `ess-sas-submit-method`.

`F2` performs the '`refresh`' operation on the current buffer. '`refresh`' compares the buffer's last modified date/time with the file's last modified date/time and replaces the buffer with the file if the file is newer. This is the same operation that is automatically performed when `LOG`, `OUTPUT`, '`filetype-1`' or `F11` are pressed.

'`filetype-1`' switches you to a file with the same file name as your `.sas` file, but with a different extension (`.txt` by default) and performs '`refresh`'. You can over-ride the default extension; for example in your `~/.emacs` or `~/.xemacs/init.el` file:

```
(setq ess-sas-suffix-1 "csv") ; for example
```

`F9` will prompt you for the name of a permanent SAS dataset near point to be opened for viewing by `PROC FSEDIT`. You can control the SAS batch command-line with `ess-sas-data-view-submit-options`. For controlling the SAS batch commands, you have the global variables `ess-sas-data-view-libname` and `ess-sas-data-view-fsview-command` as well as the buffer-local variable `ess-sas-data-view-fsview-statement`. If you have your SAS `LIBNAME` defined in `~/autoexec.sas`, then the defaults for these variables should be sufficient.

Similarly, *C-F9* will prompt you for the name of a permanent SAS dataset near point to be opened for viewing by `PROC INSIGHT`. You can control the SAS batch command-line with `ess-sas-data-view-submit-options`. For controlling the SAS batch commands, you have the global variables `ess-sas-data-view-libname` and `ess-sas-data-view-insight-command` as well as the buffer-local variable `ess-sas-data-view-insight-statement`.

`F10` toggles ESS[SAS] mode for `.log` files which is off by default (technically, it is `SAS-log-mode`, but it looks the same). The syntax highlighting can be helpful in certain debugging situations, but large `.log` files may take a long time to highlight.

`F11` is the same as '`filetype-1`' except it is `.dat` by default.

`F12` will prompt you for the name of a `GSASFILE` near the point in `.log` to be opened for viewing either with emacs or with an external viewer. Depending on your version of emacs and the operating system you are using, emacs may support `.gif` and `.jpg` files internally. You may need to change the following variables for your own situation. `ess-sas-graph-view-suffix-regexp` is a regular expression of supported file types defined via file name extensions. `ess-sas-graph-view-viewer-default` is the default external viewer for your platform. `ess-sas-graph-view-viewer-alist` is an alist of exceptions to the default; i.e. file types and their associated viewers which will be used rather than the default viewer.

```
(setq ess-sas-graph-view-suffix-regexp (concat "[.]\\([eE]?[pP][sS]\\|"
```

```
"[pP][dD][fF]\\|[gG][iI][fF]\\|[jJ][pP][eE]?[gG]\\|"
"[tT][iI][fF][fF]?\\)")) ;; default
(setq ess-sas-graph-view-viewer-default "kodakimg") ;; Windows default
(setq ess-sas-graph-view-viewer-default "sdtimage") ;; Solaris default
(setq ess-sas-graph-view-viewer-alist
   '(("[eE]?[pP][sS]" . "gv") ("[pP][dD][fF]" . "gv")) ;; default w/ gv
```

C-F2 produces US landscape by default, however, it can produce A4 landscape (first line for "global" key mapping, second for "local"):

```
(global-set-key [(control f2)] 'ess-sas-rtf-a4-landscape)
(define-key sas-mode-local-map [(control f2)] 'ess-sas-rtf-a4-landscape)
```

13.6 iESS[SAS]–Interactive SAS processes

Inferior ESS (iESS) is the method for interfacing with interactive statistical processes (programs). iESS[SAS] is what is needed for interactive SAS programming. iESS[SAS] works best with the SAS command-line option settings "-stdio -linesize 80 -noovp -nosyntaxcheck" (the default of **inferior-SAS-args**).

-stdio
> required to make the redirection of stdio work

-linesize 80
> keeps output lines from folding on standard terminals

-noovp
> prevents error messages from printing 3 times

-nosyntaxcheck
> permits recovery after syntax errors

To start up iESS[SAS] mode, use:

M-x SAS

The *SAS:1.log* buffer in **ESStr** mode corresponds to the file *foo.log* in SAS batch usage and to the 'SAS: LOG' window in the SAS Display Manager. All commands submitted to SAS, informative messages, warnings, and errors appear here.

The *SAS:1.lst* buffer in **ESSlst** mode corresponds to the file *foo.lst* in SAS batch usage and to the 'SAS: OUTPUT' window in the SAS Display Manager. All printed output appears in this window.

The *SAS:1* buffer exists solely as a communications buffer. The user should never use this buffer directly. Files are edited in the *foo.sas* buffer. The *C-c C-r* key in ESS[SAS] is the functional equivalent of bringing a file into the 'SAS: PROGRAM EDITOR' window followed by SUBMIT.

For example, open the file you want to work with.

C-x C-f foo.sas

foo.sas will be in ESS[SAS] mode. Edit as appropriate, and then start up SAS with the cursor in the *foo.sas* buffer.

M-x SAS

Four buffers will appear on screen:

Buffer	Mode	Description
`foo.sas`	`ESS[SAS]`	your source file
`*SAS:1*`	`iESS[SAS:1]`	iESS communication buffer
`*SAS:1.log*`	`Shell ESStr []`	SAS log information
`*SAS:1.lst*`	`Shell ESSlst []`	SAS listing information

If you would prefer each of the four buffers to appear in its own individual frame, you can arrange for that. Place the cursor in the buffer displaying `foo.sas`. Enter the sequence `C-c C-w`. The cursor will normally be in buffer `foo.sas`. If not, put it there and `C-x b foo.sas`.

Send regions, lines, or the entire file contents to SAS (regions are most useful: a highlighted region will normally begin with the keywords `DATA` or `PROC` and end with `RUN;`), `C-c C-r`.

Information appears in the log buffer, analysis results in the listing buffer. In case of errors, make the corrections in the `foo.sas` buffer and resubmit with another `C-c C-r`.

At the end of the session you may save the log and listing buffers with the usual `C-x C-s` commands. You will be prompted for a file name. Typically, the names `foo.log` and `foo.lst` will be used. You will almost certainly want to edit the saved files before including them in a report. The files are read-only by default. You can make them writable by the emacs command `C-x C-q`.

At the end of the session, the input file `foo.sas` will typically have been revised. You can save it. It can be used later as the beginning of another iESS[SAS] session. It can also be used as a batch input file to SAS.

The `*SAS:1*` buffer is strictly for ESS use. The user should never need to read it or write to it. Refer to the `.lst` and `.log` buffers for monitoring output!

Troubleshooting: See Section 13.7 [iESS(SAS)–Common problems], page 75.

13.7 iESS[SAS]–Common problems

1. iESS[SAS] does not work on Windows. In order to run SAS inside an emacs buffer, it is necessary to start SAS with the `-stdio` option. SAS does not support the `-stdio` option on Windows.

2. If `M-x SAS` gives errors upon startup, check the following:

   - you are running Windows: see 1.
   - `ess-sas-sh-command` (from the ESS `etc` directory) needs to be executable; too check, type `M-x dired`; if not, fix it as follows, type `M-:`, then at the minibuffer prompt 'Eval:', type `(set-file-modes "ess-sas-sh-command" 493)`.
   - `sas` isn't in your executable path; to verify, type `M-:` and at the minibuffer prompt 'Eval:', type `(executable-find "sas")`

3. `M-x SAS` starts SAS Display Manager. Probably, the command `sas` on your system calls a shell script. In that case you will need to locate the real `sas` executable and link to it. You can execute the UNIX command:

   ```
   find / -name sas -print
   ```

   Now place a soft link to the real `sas` executable in your `~/bin` directory, with for example

```
cd ~/bin
ln -s /usr/local/sas9/sas sas
```

Check your PATH environment variable to confirm that ~/bin appears before the directory in which the sas shell script appears.

13.8 ESS[SAS]–Graphics

Output from a SAS/GRAPH PROC can be displayed in a SAS/GRAPH window for SAS batch on Windows or for both SAS batch and interactive with XWindows on UNIX. If you need to create graphics files and view them with F12, then include the following (either in *foo*.sas or in ~/autoexec.sas):

```
filename gsasfile 'graphics.ps';
goptions device=ps gsfname=gsasfile gsfmode=append;
```

PROC PLOT graphs can be viewed in the listing buffer. You may wish to control the vertical spacing to allow the entire plot to be visible on screen, for example:

```
proc plot;
    plot a*b / vpos=25;
run;
```

13.9 ESS[SAS]–Windows

- iESS[SAS] does not work on Windows. See Section 13.7 [iESS(SAS)–Common problems], page 75.

- ESS[SAS] mode for editing SAS language files works very well. See Section 13.2 [ESS(SAS)–Editing files], page 68.

- There are two execution options for SAS on Windows. You can use batch. See Section 13.4 [ESS(SAS)–Batch SAS processes], page 69.

  Or you can mark regions with the mouse and submit the code with 'submit-region' or paste them into SAS Display Manager.

14 ESS for BUGS

ESS[BUGS] was originally designed for use with BUGS software. Later, it evolved to support JAGS as a dialect of the BUGS language via ESS[JAGS], however, ESS[JAGS] is documented in the section Help for JAGS. ESS[BUGS] provides 5 features. First, BUGS syntax is described to allow for proper fontification of statements, distributions, functions, commands and comments in BUGS model files, command files and log files. Second, ESS creates templates for the command file from the model file so that a BUGS batch process can be defined by a single file. Third, ESS provides a BUGS batch script that allows ESS to set BUGS batch parameters. Fourth, key sequences are defined to create a command file and submit a BUGS batch process. Lastly, interactive submission of BUGS commands is also supported.

14.1 ESS[BUGS]–Model files

Model files with the `.bug` extension are edited in ESS[BUGS] mode if (`require 'ess-bugs-d`) was performed. Model files with the `.jag` extension are edited in ESS[JAGS] mode if (`require 'ess-jags-d`) was performed. Three keys are bound for your use in ESS[BUGS], *F2*, *C-c C-c* and *=*. *F2* performs the same action as it does in ESS[SAS], See Section 13.5 [ESS(SAS)–Function keys for batch processing], page 71. *C-c C-c* performs the function `ess-bugs-next-action` which you will use a lot. Pressing it in an empty buffer for a model file will produce a template for you. *=* inserts the set operator, `<-`.

14.2 ESS[BUGS]–Command files

To avoid extension name collision, `.bmd` is used for BUGS command files and `.jmd` for JAGS command files. When you have finished editing your model file and press *C-c C-c*, a command file is created if one does not already exist. When you are finished editing your command file, pressing *C-c C-c* again will submit your command file as a batch job.

14.3 ESS[BUGS]–Log files

To avoid extension name collision, `.bog` is used for BUGS log files. The command line generated by ESS creates the `.bog` transcript file. Similarly, `.jog` is used for JAGS log files.

15 ESS for JAGS

ESS[BUGS] was originally designed for use with BUGS software. Later, it evolved to support JAGS as a dialect of the BUGS language via ESS[JAGS]. Since BUGS and JAGS are quite similar, ESS[BUGS] and ESS[JAGS] are necessarily similar. ESS[JAGS] provides 4 features. First, JAGS syntax is described to allow for proper fontification of statements, distributions, functions, commands and comments in JAGS model files, command files and log files. Second, ESS creates templates for the command file from the model file so that a JAGS batch process can be defined by a single file. Third, ESS provides a JAGS batch script that allows ESS to set JAGS batch parameters. Fourth, key sequences are defined to create a command file and submit a JAGS batch process.

15.1 ESS[JAGS]–Model files

Model files with the `.bug` extension are edited in ESS[BUGS] mode if (require 'ess-bugs-d) was performed or edited in ESS[JAGS] mode if (require 'ess-jags-d). Three keys are bound for your use in ESS[BUGS], *F2*, *C-c C-c* and *=*. *F2* performs the same action as it does in ESS[SAS], See Section 13.5 [ESS(SAS)–Function keys for batch processing], page 71. *C-c C-c* performs the function `ess-bugs-next-action` which you will use a lot. Pressing it in an empty buffer for a model file will produce a template for you. *=* inserts the set operator, `<-`.

ESS[JAGS] does not support "replacement" variables which were part of ESS[BUGS]. Although, "replacement" variables are an extremely convenient feature, creating the elisp code for their operation was challenging. So, a more elisp-ish approach was adopted for ESS[JAGS]: elisp local variables. These variables are created as part of the template, i.e. with the first press of *C-c C-c* in an empty buffer. For your `.bug` file, they are named `ess-jags-chains`, `ess-jags-monitor`, `ess-jags-thin`, `ess-jags-burnin` and `ess-jags-update`; they appear in the `Local Variables` section. When you are finished editing your model file, pressing *C-c C-c* will perform the necessary actions to build your command file for you.

The `ess-jags-chains` variable is the number of chains that you want to initialize and sample from; defaults to 1. The `ess-jags-monitor` variable is a list of variables that you want monitored: encase each variable in double quotes. When you press *C-c C-c*, the appropriate statements are created in the command file to monitor the list of variables. By default, no variables are explicitly monitored which means JAGS will implicitly monitor all "default" variables. The `ess-jags-thin` variable is the thinning parameter. By default, the thinning parameter is set to 1, i.e. no thinning. The `ess-jags-burnin` variable is the number of initial samples to discard. By default, the burnin parameter is set to 10000. The `ess-jags-update` variable is the number of post-burnin samples to keep. By default, the update parameter is set to 10000. Both `ess-jags-burnin` and `ess-jags-update` are multiplied by `ess-jags-thin` since JAGS does not do it automatically.

15.2 ESS[JAGS]–Command files

To avoid extension name collision, `.bmd` is used for BUGS command files and `.jmd` for JAGS command files. For your `.jmd` file, there is only one variable, `ess-jags-command`, in the `Local Variables` section. When you have finished editing your model file and press *C-c*

C-c, a command file is created if one does not already exist. When you are finished editing your command file, pressing *C-c C-c* again will submit your command file as a batch job. The `ess-jags-command` variable allows you to specify a different JAGS program to use to run your model; defaults to "jags".

15.3 ESS[JAGS]–Log files

The `.jog` extension is used for JAGS log files. You may find *F2* useful to refresh the `.jog` if the batch process over-writes or appends it.

16 Bugs and Bug Reporting, Mailing Lists

16.1 Bugs

- Commands like `ess-display-help-on-object` and list completion cannot be used while the user is entering a multi-line command. The only real fix in this situation is to use another ESS process.

- The `ess-eval-` commands can leave point in the ESS process buffer in the wrong place when point is at the same position as the last process output. This proves difficult to fix, in general, as we need to consider all *windows* with `window-point` at the right place.

- It's possible to clear the modification flag (say, by saving the buffer) with the edit buffer not having been loaded into S.

- Backup files can sometimes be left behind, even when `ess-keep-dump-files` is `nil`.

- Passing an incomplete S expression to `ess-execute` causes ESS to hang.

- The function-based commands don't always work as expected on functions whose body is not a parenthesized or compound expression, and don't even recognize anonymous functions (i.e. functions not assigned to any variable).

- Multi-line commands could be handled better by the command history mechanism.

- Changes to the continuation prompt in R (e.g. `options(continue = " ")`) are not automatically detected by ESS. Hence, for now, the best thing is not to change the continuation prompt. If you do change the continuation prompt, you will need to change accordingly the value of `inferior-ess-secondary-prompt` in `R-customize-alist`.

16.2 Reporting Bugs

Please send bug reports, suggestions etc. to ESS-bugs@stat.math.ethz.ch

The easiest way to do this is within Emacs by typing

`M-x ess-submit-bug-report`

This also gives the maintainers valuable information about your installation which may help us to identify or even fix the bug.

If Emacs reports an error, backtraces can help us debug the problem. Type "M-x set-variable RET debug-on-error RET t RET". Then run the command that causes the error and you should see a *Backtrace* buffer containing debug information; send us that buffer.

Note that comments, suggestions, words of praise and large cash donations are also more than welcome.

16.3 Mailing Lists

There is a mailing list for discussions and announcements relating to ESS. Join the list by sending an e-mail with "subscribe ess-help" (or "help") in the body to ess-help-request@stat.math.ethz.ch; contributions to the list may be mailed to ess-help@stat.math.ethz.ch. Rest assured, this is a fairly low-volume mailing list.

The purposes of the mailing list include

- helping users of ESS to get along with it.
- discussing aspects of using ESS on Emacs and XEmacs.
- suggestions for improvements.
- announcements of new releases of ESS.
- posting small patches to ESS.

16.4 Help with emacs

Emacs is a complex editor with many abilities that we do not have space to describe here. If you have problems with other features of emacs (e.g. printing), there are several sources to consult, including the emacs FAQs (try *M-x xemacs-www-faq* or *M-x view-emacs-FAQ*) and EmacsWiki (http://www.emacswiki.org). Please consult them before asking on the mailing list about issues that are not specific to ESS.

Appendix A Customizing ESS

ESS can be easily customized to your taste simply by including the appropriate lines in your `.emacs` file. There are numerous variables which affect the behavior of ESS in certain situations which can be modified to your liking. Keybindings may be set or changed to your preferences, and for per-buffer customizations hooks are also available.

Most of these variables can be viewed and set using the Custom facility within Emacs. Type `M-x customize-group RET ess RET` to see all the ESS variables that can be customized. Variables are grouped by subject to make it easy to find related variables.

Indices

Key index

,
, ... 56

|
{ 39
} 39

C
C-c 39
C-c ' 32, 38
C-c C-. a 60
C-c C-. d 60
C-c C-. H 60
C-c C-. l 60
C-c C-. m 60
C-c C-. o 60
C-c C-. r 60
C-c C-. s 60
C-c C-. u 60
C-c C-. w 60
C-C C-b 35
C-c C-c 33
C-C C-c 34
C-c C-e C-d 37
C-C C-f 34
C-C C-j 34
C-c C-l 32
C-c C-o C-c 52
C-c C-o C-h 53
C-c C-o C-o 52
C-c C-o C-r 52
C-c C-o C-t 53
C-c C-o n 53
C-c C-o p 53
C-c C-q 32
C-C C-r 34
C-c C-s 32
C-c C-t C-a 54
C-c C-t C-r 54
C-c C-t C-t 54

C-c C-t l 54
C-c C-v 32
C-c C-w 36
C-c C-x 31
C-c C-z 33
C-C M-b 35
C-C M-f 34
C-C M-j 34
C-C M-r 34
C-c RET 36
C-j 39
C-M-a 40
C-M-e 40
C-M-h 40
C-M-q 39
C-M-x 34
C-x ' 32

E
ESC C-a 40
ESC C-e 40
ESC C-h 40
ESC C-q 39

M
M-; 39
M-? 46
M-C-q 39
M-n l 61
M-n P 61
M-n s 61
M-RET 36

R
RET 26, 36

T
TAB 38

Function and program index

B
backward-kill-word 26

C
comint-backward-matching-input 27
comint-bol 26
comint-copy-old-input 28